In *Situ* Hybridization

Applications to Neurobiology

In Situ Hybridization

Applications to Neurobiology

Edited by

Karen L. Valentino

James H. Eberwine

Jack D. Barchas

New York   Oxford
OXFORD UNIVERSITY PRESS
1987

Oxford University Press

Oxford   New York   Toronto
Delhi   Bombay   Calcutta   Madras   Karachi
Petaling Jaya   Singapore   Hong Kong   Tokyo
Nairobi   Dar es Salaam   Cape Town
Melbourne   Auckland

and associated companies in
Beirut   Berlin   Ibadan   Nicosia

Library of Congress Cataloging-in-Publication Data
In situ hybridization.
Bibliography: p.   Includes index.
1. Neurochemistry—Technique.   2. Nucleic acid hybridization.
3. Gene expression.   4. Neurobiology—Technique.
I. Valentino, Karen L.   II. Eberwine, James H.
III. Barchas, Jack D., 1935–
QP356.3.I49   1987   616.89'07   86-23721
ISBN 0-19-504859-8

Printing (last digit): 9 8 7 6 5 4 3 2

Printed in the United States of America
on acid-free paper

Foreword

One of the cultural clichés that has often been relevant during my career is Alexander Pope's maxim: "Be not the first to set the old aside, nor yet the last to leave the new untried." This is probably a practical principle for anyone seeking new strategies to sort out the spatial distribution of specific molecules in specific cells and tissues. Neurobiologists have made concerns over spatial distribution into a major art form known as chemical neuroanatomy, which constantly requires new methods to sort out the chemical differences between the distinguishable "types" of cells in the nervous system as one step in determining how cells are connected and function.

As the newest method of getting such chemical information, *in situ* hybridization is experiencing the same sorts of growing pains that were observed with earlier waves of chemical dissection of the brain. The "classic" staining era revealed important facts about neuronal and glial morphology; the basis for most of the staining of either healthy or dying cells was entirely unknown, however, and the results were highly variable, capricious, and open to substantial subjective interpretation. The era of specific staining, which began with the development of substrates for enzymes that could lead to insoluble light or electron-dense reaction products, attracted me to enzyme histochemistry of the brain when it seemed impossible that we would ever be able to identify specific neurotransmitters directly. Those methods were very attractive for what they might tell us, but fraught with artifacts (remember "nothing dehydrogenase"?), and very labor-intensive as well. To see an enzyme react in a tissue section required walking a fine line between enough fixative to keep the cell's structure recognizable, but not so much as to kill the enzyme.

Some transmitter-related enzymes were seen—cholinesterases, ATPases, and monoamine oxidases—but these were really fairly remote from the more direct molecular markers of neurons using a specific transmitter. Although some presumptive monoamine and amino acid neurons were detectable by both light and electron microscopic autoradiography through their ability to accumulate radioactive transmitter, this method had rather limited applicability, required many controls for clean interpretation, and fell definitely into the artform class of scientific methods.

Soon thereafter came the era of direct visualization of those elusive small transmitter molecules. The techniques of freeze-dry, formaldehyde-induced fluorescence microscopy were dramatic in their intensity, sensitivity, and clarity. They were also extraordinarily hard to do consistently unless one was blessed with long, arid winters and enjoyed the constant sound of vacuum pumps on freeze driers. Very few laboratories were able to practice this artform with the regularity of the Swedish groups at Karolinska and Lund, and it took more than a decade to devise means for standardizing these methods and avoiding the need to freeze unfixed tissue. Furthermore, the fluorescence aldehyde approach was limited to the two catecholamines, dopamine and norepinephrine, and to transient visualization of 5-hydroxytryptamine.

That era wound down rapidly with anatomic exploitation of the ideal cytochemical reagent—namely, an antibody that could read the specific molecules one wanted to see. First came the catecholamine-synthesizing enzyme, dopamine beta-hydroxylase, which was abundant and already close to purification. This was quickly followed by antibodies to other relevant purified proteins and, later, when proper bridging methods had been worked out, to smaller and smaller peptides and eventually to the amines and the amino acids themselves—thus, in principle to almost anything a lymphocyte can distinguish, which turns out to be almost every molecule made. The ability to make monoclonal antibodies against these markers, thereby restricting the number of epitopes being detected, gave this immunocytochemical approach several additional important advantages.

Immunocytochemical methods have, in the past decade, become the major means of identifying the location of specific molecular markers. Perhaps the most widely practiced histochemical art, immunochemistry has the important attribute of using the same reagent for both structural and chemical analyses. As with earlier approaches, one must be aware of the possible artifacts, interpret the data with caution and at times even skepticism, and continue to walk the tightrope between underfixation and loss of cellular structure and overfixation and loss of immunoaffinity for the antibody. Furthermore, with immunocytochemistry it is at last possible to use methods of comparable sensitivity for neuronal markers of all kinds, not simply transmitters or transmitter-related markers such as synthetic and catabolic enzymes and receptors. In fact, any component that can be even partially purified can be used to raise antisera, which then become tools for further purification and better antisera. Clearly, tools such as antisera permit chemists and cytologists to collaborate. Both can use the same reagent in their preferred reaction systems. Their data can then interrelate, and the resultant complementary data sets lend themselves to rapid exploitation and widespread applications.

As this important first book on the application of *in situ* hybridization methods to neurobiology makes clear, we now have at hand a means to probe the very core of the cell's chemical identity by constructing equally ideal reagents of extraordinary specificity and sensitivity made by vectors, bacteria, or gene machines that can read the genes or messenger RNAs, by which cells express their unique population of structural and functional proteins. The stable strong signals, and the ability to combine *in situ* methods with many other methods of visualizing spatially the same chemical components of cells, promise to be very useful. These methods are especially promising for the chemically heterogeneous sets of morphologically similar neurons and glia in nervous systems. Like their methodological predecessors, most of the possible artifacts and limitations are still probably to be discovered, and a large number of delicate steps and potential pitfalls are still to be conquered before widespread applications can be envisioned. Nevertheless, I share the enthusiasm of Drs. Valentino, Eberwine, and Barchas for the potential opportunities this new methodology affords and applaud their foresight in seeing the need for this volume and compiling it with such a fine array of pioneering participants. This time it is unnecessary to set the old aside to try the new, since this new method fully complements all of the methods that have gone before. I welcome this new arrow in our cytochemical quiver.

Floyd E. Bloom

Preface

In situ hybridization has enormous potential importance for neurobiology and research in the clinical neurosciences, including psychiatry. The method, which is derived from techniques in molecular biology and anatomy, involves the use of gene-based probes to study changes in genetic expression in specific cells of brain and other tissues.

The basis of *in situ* hybridization is that a piece of labeled ribonucleic acid (RNA) or complementary deoxyribonucleic acid (cDNA) from a gene of interest can be hybridized to a section of tissue, permitting determination of the cells transcribing that particular gene. Over the past decade and a half, the method has proven very powerful for studies of gene transcription and regulation. It has begun to be applied to the central nervous system only in the past few years.

The methodology of *in situ* hybridization has now advanced to the point of being sensitive and practical. As a methodology, it can provide more exact and sensitive information than any simple chemical method for measurement of genetic expression; it has the advantage of great specificity in terms of both the genetic information expressed and identification of the cells in which that information is expressed.

Because the applications are so new, there are virtually no reference works for investigators interested in working in this field. *In situ* hybridization is a difficult technique to establish experimentally, and many variables must be controlled before one can obtain reproducible data. Although a few laboratories that have pioneered the method have begun asking interesting biological questions using *in situ* hybridization, many other scientists initiating *in situ* studies have had difficulty dealing with the practical aspects of the technique. Immunohistochemistry revolutionized neurobiology and is now a standard tool in advancing neuroscience. We think *in situ* hybridization awaits a similar future.

To address the problems associated with this new methodology, the MacArthur Foundation Network on the Psychobiology of Depression and Other Affective Disorders convened a conference on the subject. Many scholars who have contributed the most to the development of the technique participated in the symposium.

Because of the importance of technical issues for the development and application of this approach, particular attention was paid to issues of methodology. The goal was to examine the methodology in a more rigorous manner to see if it could be codified, while permitting exploration of the technique's current potential and limits. It was hoped that such an exploration would encourage scientists to use the method and help them to establish the technique in their laboratories.

The conference revealed a surprising degree of consensus on technical issues. This monograph reflects the conference, but the papers have been expanded to give a full and detailed sense of the methodology and some of its potential roles.

The invited scholars included neurobiologists and others with different perspectives on the technique, reflecting the broad range of uses of the method throughout biology. Thus, some chapters deal with general biological approaches and others with more specific applications of *in situ* hybridization to neurobiology.

Floyd Bloom, who has been active in this area, and who also coauthored a chapter, has graciously provided the Foreword. An overview chapter has been included to provide a background and rationale for the technique, as well as to consider some of the reasons for variation in methods between laboratories.

The contributors and their areas of interest include the following:

John Coghlan of the Howard Florey Institute in Australia has achieved remarkable success using oligonucleotide probes for a wide variety of peptide and hormone messenger RNAs. He and his colleagues have developed the technique of whole embryo *in situ* hybridization such that it is possible to examine every tissue in the developing fetus for the expression of any messenger RNA for which there is a probe.

Robert Angerer, of the Department of Biology at the University of Rochester, was one of the pioneers in the use of *in situ* hybridization. From work in his laboratory dealing with sea urchin embryo development has come much of the practical methodology for *in situ* hybridization using RNA probes and striking demonstrations of the success of the technique for studying basic biological problems. He has applied the theory of solution hybridization (three-dimensional) to *in situ* hybridization (two-dimensional). This has produced a high degree of reproducibility in his *in situ* studies.

Robert Singer of the University of Massachusetts has made valuable contributions through his use of biotinylated probes rather than radiolabeled probes. The continued development of this approach will obviate the use of radioactivity in this technique. Additionally, biotinylated probes used in conjunction with fluorescent- (or other) labeled antibodies have the potential of further increasing the sensitivity of the *in situ* technique.

Sue Griffin of the University of Arkansas has been particularly concerned with formulating some of the methodology as it applies to studies of the central nervous system. She has used the technique to examine actin and tubulin gene expression in rat brain development as in animals with graft-versus-host disease.

Beth Schachter of the Mt. Sinai Medical Center was instrumental in application of the *in situ* technique to mammalian brain. She has worked with probes for various hormones and has used the technique to perform studies of steroid regulation of neuropeptide gene expression.

Stanley Watson of the University of Michigan and his collaborators were the first to report results using this technique in the mammalian brain. He has used the *in situ* method to localize neuroregulators using oligonucleotide and RNA probes. These authors have also addressed the problem of doing *in situ* hybridization and immunocytochemistry simultaneously, utilizing their splendid sense of histochemical procedures.

Gerald Higgins of the Scripps Clinic provides a powerful example of the use of the technique to advance neurobiology through studies involving identification of unique or cell-type and system-specific neuronal mRNAs.

Robert Milner of the Scripps Clinic has been involved in using the method to analyze brain-specific probes he and his collaborators have isolated. The *in situ* technique has permitted the identification of new, previously unrecognized neurotransmitters and neuromodulators, thus enhancing our knowledge of neuroactive agents and the systems of which they are a part.

John Pintar of Columbia has been engaged in a series of developmental studies dealing with control of pituitary hormone gene expression. These studies have demonstrated the possibility of performing important biological investigations in embryos.

Ashley Haase of the University of Minnesota was one of the first investigators to explore the *in situ* technique for detection of viral DNA. He has developed what is thought to be the most sensitive *in situ* system available at this time and provides an example of the power of the technique, using elegant studies of viral infections.

Since a number of common threads emerged from the question-and-answer sessions following each talk, it made sense to combine these discussions into a single section. This is done in the Appendix.

Edward Jones and collaborators at Irvine are using *in situ* hybridization to study function in primate cerebral cortex and have presented some of their preliminary data in Appendix B.

In developing this monograph, we asked the authors to stress their methods, so the readers can use the various chapters to establish their own experimental protocols. There are many important choices to be made in undertaking *in situ* hybridization, including the selection of an appropriate probe to hybridize to the corresponding RNA. The choices range from the use of single- or double-stranded RNA or DNA as a probe, to whether the probe should be labeled with radioactivity—and if so, with which isotope—to the use of biotin as a label. The authors have faced these and other problems in different ways. Their detailed resolutions, which are presented here, may be helpful to the reader establishing these methods. At first glance, many of the approaches will appear similar, yet as anyone attempting *in situ* studies can attest, the subtle differences are critically important.

In preparing the volume we considered terms other than "*in situ* hybridization" for the technique, including "insitu hybridization." Dr. Coghlan and his colleagues proposed "hybridization histochemistry," which has the advantage of distinguishing this technique from that used in chromosomal mapping studies. Terminology that distinguishes between various forms of *in situ* hybridization concerned with identifying messenger RNA or with identifying infectious viral DNA would also be useful. In this case, we chose the traditional approach, but we raise the issue of finding a less cumbersome and more informative convention for our colleagues in the scientific community to consider. We believe a new format is desirable.

We have many people to thank for their help with the conference and monograph. The conference was sponsored by the MacArthur Foundation Network and Stanford's Nancy Pritzker Laboratory of Behavioral Neurochemistry. The MacArthur Foundation has established a series of preformed networks, consisting of scientists it has selected, to link research in certain areas of mental health and illness. The Psychobiology of Depression Network has a set of nodes, each directed by a research investigator. It includes the departments of psychiatry at Harvard (Joseph Lipinski), Illinois (John Davis), Pittsburgh (David Kupfer), and Stanford (Philip Berger), as well as the department of human genetics at Yale (Kenneth Kidd) and the laboratory of preclinical neuroscience at the Scripps Clinic (Floyd Bloom).

The goal of the Psychobiology of Depression Network has been to encourage clinical research as well as basic studies, many of which ultimately may permit collaborative investigations either within or outside of the network. A stated goal of the foundation, as well as the network, has been to encourage communication between scientists. The network has attempted to facilitate efforts within its disciplines through the organization of conferences. Several have been completed and are in various stages of publication. They will result in monographs dealing with circadian rhythms, childhood depression, learning and affective mechanisms, and sociophysiology.

The importance of basic research, and its role in relation to the study of mental illness, is one of the bases on which the network was organized. It was the ultimate rationale for this conference.

The foundation was most generous. It provided funds for the meeting and also supplied resources for color reproduction of some of the graphics, an important addition in a volume based on a visual methodology. The color micrographs should be a considerable aid to investigators attempting to establish the methodology.

We appreciate the support of the foundation as well as the encouragement of David Kupfer, professor and chairman of the department of psychiatry at the University of Pittsburgh, who currently serves as coordinator of the network.

Special efforts by several individuals within the Nancy Pritzker Laboratory contributed to the success of this endeavor. Cathy Dolton was tireless in making arrangements for the speakers and in dealing with the mechanics of the conference. Edna Dorles performed her usual magic with budgetary issues. Alicia Fritchle was our sound engineer and also transcribed the discussion from each talk. She performed the heroic task of retyping the entire manuscript in a computerized format under extremely tight time constraints. Other members of the laboratory to whom we are indebted for helping with the meeting or with preparation of the monograph were Richard Dorin, Mark Schaefer, Larry Tecott, Irene Inman, Joan Hunter, George Makk and Ines Zangger. The index was prepared by Steven Sorensen.

A special debt of gratitude is owed to Oxford University Press for their help and encouragement of this project. The editor at Oxford University Press with whom we worked was William Frucht. Jeffrey House, senior vice-president for Oxford, is a person with special sensitivity to the publishing needs in the neurosciences. He was, as always, encouraging.

Finally, the participants at the symposium—speakers and guests—aided the focus of the conference by dealing with comparative details of the methodology and asking direct and valuable questions. We owe an enormous debt of gratitude to all of those who prepared extensive, method-oriented chapters in a coordinated, time-relevant manner. We hope the volume will be of help in the field of neurobiology as the use of this methodology becomes more common and spreads to important biological and clinical questions.

Nancy Pritzker Laboratory   K. L. V.
Stanford, California   J. H. E.
                                                                                      J. D. B.

Contents

Contributors

Huda Akil
Mental Health Research Institute
University of Michigan
Ann Arbor, MI 48109

P. Aldred
Howard Florey Institute of
  Experimental Physiology and
  Medicine
University of Melbourne
Parkville, Victoria
3052 Australia

Lynne M. Angerer
Department of Biology
University of Rochester
Rochester, NY 14627

Robert C. Angerer
Department of Biology
University of Rochester
Rochester, NY 14627

Jack D. Barchas
Nancy Pritzker Laboratory of
  Behavioral Neurochemistry
Department of Psychiatry and
  Behavioral Science
Stanford University School of
  Medicine
Stanford, CA 94305

Floyd E. Bloom
Division of Preclinical Neuroscience
  and Endocrinology
Research Institute of Scripps Clinic
10666 North Torrey Pines Road
La Jolla, CA 92037

Ralph A. Bradshaw
Department of Biological Chemistry
University of California
Irvine, CA 92717

Sharon Burke
Mental Health Research Institute
University of Michigan
Ann Arbor, MI 48109

John P. Coghlan
Howard Florey Institute of
  Experimental Physiology and
  Medicine
University of Melbourne
Parkville, Victoria
3052 Australia

James H. Eberwine
Nancy Pritzker Laboratory of
  Behavioral Neurochemistry
Department of Psychiatry and
  Behavioral Science
Stanford University School of
  Medicine
Stanford, CA 94305

J. R. Fraser
Department of Medicine
Royal Melbourne Hospital
Royal Parade Parkville
Victoria
3052 Australia

W. Sue T. Griffin
University of Arkansas for Medical
  Sciences
4301 W. Markham
Little Rock, AR 72205

Ashley T. Haase
Department of Microbiology
University of Minnesota
420 Delaware Street S.E.
Minneapolis, MN 55455

J. Haralambidis
Howard Florey Institute of
  Experimental Physiology and
  Medicine
University of Melbourne
Parkville, Victoria
3052 Australia

Stewart H. C. Hendry
Department of Anatomy and
  Neurobiology
University of California
Irvine, CA 92717

Gerald A. Higgins
Department of Molecular Biology
Research Institute of Scripps Clinic
10666 Torrey Pines Road
La Jolla, CA 92037

P. Isackson
Department of Biological Chemistry
University of California
Irvine, CA 92717

Edward G. Jones
Department of Anatomy and
  Neurobiology
University of California
Irvine, CA 92717

Jeffrey E. Kelsey
Mental Health Research Institute
University of Michigan
Ann Arbor, MI 48109

Jeanne Bentley Lawrence
Department of Anatomy
University of Massachusetts Medical
  School
55 Lake Avenue North
Worcester, MA 01605

Delia Ines Lugo
Department of Anatomy and Cell
  Biology
Columbia University

College of Physicians and Surgeons
630 West 168 Street
New York, NY 10032

Robert J. Milner
Division of Preclinical Neuroscience
  and Endocrinology
Research Institute of Scripps Clinic
10666 North Torrey Pines Road
La Jolla, CA 92037

Jennifer D. Penschow
Howard Florey Institute of
  Experimental Physiology and
  Medicine
University of Melbourne
Parkville, Victoria
3052 Australia

John E. Pintar
Department of Anatomy and Cell
  Biology
Columbia University
College of Physicians and Surgeons
630 West 168 Street
New York, NY 10032

Roya N. Rashtchian
Department of Anatomy
University of Massachusetts Medical
  School
55 Lake Avenue North
Worcester, MA 01605

Beth S. Schachter
Department of Obstetrics, Gynecology,
  and Reproductive Science and
  Department of Anatomy
Mount Sinai School of Medicine
New York, NY 10029

Hartwig Schmale
Institut für Zellbiochemie und
  Klinische und Neurobiologie
Universitat Hamburg
Martinistrasse 52/2
Hamburg 20
Federal Republic of Germany

Thomas G. Sherman
Mental Health Research Institute
University of Michigan
Ann Arbor, MI 48109

Robert H. Singer
Deparment of Anatomy
University of Massachusetts Medical
  School
55 Lake Avenue North
Worcester, MA 01605

Mark H. Stoler
Department of Pathology
University of Rochester
Rochester, NY 14627

Laurence H. Tecott
Nancy Pritzker Laboratory of
  Behavioral Neurochemistry
Department of Psychiatry and
  Behavioral Science
Stanford University School of
  Medicine
Stanford, CA 94305

G. W. Tregear
Howard Florey Institute of
  Experimental Physiology and
  Medicine
University of Melbourne
Parkville, Victoria
3052 Australia

Karen L. Valentino
Nancy Pritzker Laboratory of
  Behavioral Neurochemistry
Department of Psychiatry and
  Behavioral Science
Stanford University School of
  Medicine
Stanford, CA 94305

Stanley J. Watson
Mental Health Research Institute
University of Michigan
Ann Arbor, MI 48109

Michael C. Wilson
Department of Molecular Biology
Research Institute of Scripps Clinic
10666 North Torrey Pines Road
La Jolla, CA 92037

In Situ Hybridization

Applications to Neurobiology

1

Methodological Considerations in the Utilization of *In Situ* Hybridization

Laurence H. Tecott, James H. Eberwine,
Jack D. Barchas, and Karen L. Valentino

In situ hybridization is a technique that is generating a great deal of interest among neurobiologists and others interested in genetics, virology, developmental biology, and related disciplines. This method enables the precise localization and identification of individual cells that contain a specific nucleic acid sequence, in a manner analogous to the immunocytochemical localization of cells containing a particular protein. It involves the hybridization of a nucleic acid probe with a specific target nucleic acid within a tissue section. The technique is an extension of liquid-phase hybridization methods, with which it shares many considerations. The aim of this chapter is to provide an overview of the basic technical considerations involved in *in situ* hybridization and the methods by which they are addressed.

In situ hybridization studies were initially performed to detect deoxyribonucleic acid (DNA) targets: amplified ribosomal ribonucleic acid (RNA) genes within cell nuclei (Gall and Pardue, 1969; John et al., 1969; Buongiorno-Nardelli and Amaldi, 1970). Since those early studies, much progress has been made using the technique to map the locations of particular genes within chromosomal preparations (Pardue and Dawid, 1981; Fostel et al., 1984). The development of cloning techniques has made available probes for a large and rapidly increasing number of substances. Much recent interest has focused on the use of these probes for the detection of RNA by *in situ* hybridization. Many of these RNA molecules are present in small quantities within cells, requiring methods sensitive enough to detect only a few molecules per cell. *In situ* hybridization techniques currently under development are now attaining this level of sensitivity.

There has been a wide variety of applications for *in situ* hybridization technology. The technique has been used successfully in detecting viral nucleic acid sequences within infected tissues (Brahic and Haase, 1978; Haase et al., 1984). The method is particularly useful in studying slow viruses, which may persist for long periods within infected cells and produce undetectable levels of viral antigens. In some cases, *in situ* hybridization may be the only means of detecting and localizing these viruses in tissues.

One of the most important applications of *in situ* hybridization is the detection of specific messenger RNA (mRNA) molecules. Cells make a wide variety of proteins that serve diverse functions (e.g., enzymes, structural proteins, receptors, peptide transmitters), hence they contain many different mRNA species. For example, an estimated 145,000 mRNA species can be found in mouse brain (Hahn et al., 1982), and the abundance of specific mRNA species varies greatly. Proteins involved in general processes of cell maintenance (housekeeping proteins) are present in many cell types, whereas other mRNA species may be specific to one cell type and comprise only a small fraction of the cellular mRNA content. The formidable goal of detecting such a rare mRNA *in situ* is achievable because of the specificity of complementary base pairing between a nucleic acid probe and its mRNA target.

In situ hybridization has two main advantages over current anatomical and molecular biology techniques: precise anatomical localization and great sensitivity. A commonly used method for RNA detection, Northern blotting, requires homogenization of tissue and extraction of RNA. This technique can provide neither identification of the class of cells containing the nucleic acids of interest nor their precise location. This problem is particularly important in heterogeneous tissues such as the brain, with its multitude of different cell types. In addition, the homogenization of tissues can result in a loss of sensitivity if the target is present in only a small fraction of the cells in the homogenate. For example, mRNA coding for gonadotropin-releasing hormone (GnRH) has been detected recently by *in situ* hybridization in a very small number of neurons in the rat forebrain (Shivers et al., 1986a), whereas no Northern blots detecting this mRNA have been published to date.

In some instances, mRNA levels may indicate the functional state of cells. For example, regulation of the catecholamine-synthesizing enzyme tyrosine hydroxylase is accompanied by a change in its mRNA levels (Tank et al., 1985). Regulation of secretion in some peptidergic neuroendocrine cells is also correlated with alterations in mRNA levels (Eberwine and Roberts, 1984). If an agent asserts its regulatory effect on only a small subpopulation of the cells producing a particular protein, then *in situ* hybridization may be the only available method to detect an associated change in message level. One must be careful, however, in interpreting *in situ* hybridization results alone. In many instances, mRNA levels do not correlate well with protein levels. For example, high levels of message, in the absence of detectable protein, are found for vasopressin in the hypothalamus of the Brattleboro rat (Uhl et al., 1985).

In situ hybridization complements very well the ability of standard immunohistochemical procedures to detect cellular proteins. The combined use of these techniques provides a cross-check of their specificity. If similar patterns of cellular staining are found for both methods, then the antiserum is not likely to be cross-reacting with undesired antigens, and the hybridization probe is binding specifically. *In situ* hybridization may also extend immunocytochemical results by addressing the issue of whether a positively staining cell has synthesized the detected protein or acquired it by means of uptake from the extracellular space. Detection of both protein and message in a particular cell provides strong evidence that the cell is the site of synthesis.

PRINCIPLES OF HYBRIDIZATION

Hybridization is a reaction whereby two single-stranded nucleic acid molecules recognize one another and bind by means of hydrogen bonding of complementary base pairs. DNA–DNA, RNA–RNA, and DNA–RNA duplexes all may be produced under appropriate conditions. The factors that affect hybrid stability and the kinetics of association have been well characterized for nucleic acid duplexes in solution. *In situ* hybridization is a mixed-phase reaction in which a nucleic acid probe that has been labeled with a radioisotope, or an otherwise detectable molecule hybridizes to a target nucleic acid residing in a tissue section. The immobility of the target and limitations in probe penetration of the tissue section may alter the properties of the reaction. However, the basic considerations that govern filter hybridization can aid in the selection of appropriate conditions for *in situ* hybridization.

Stability of Hybrids

Of paramount importance in determining optimal conditions for *in situ* hybridization is an understanding of the factors affecting the stability of hybrids. The two strands of a nucleic acid duplex are held together by noncovalent forces including hydrogen bonds between base pairs and hydrophobic interactions that produce an orderly stacking of planar bases, thereby minimizing their contact with water. These weak forces are easily disrupted as the temperature increases. The temperature required to disrupt these forces sufficiently for denaturation to occur varies depending on many factors. A reference point from which one can assess the stability of a hybrid is its melting temperature (T_m). This is the temperature at which half of a population of duplex molecules becomes dissociated (or "melted") into single strands. The major factors that affect the stability of DNA duplexes are related to T_m by the following equation (from Thomas and Dancis, 1973):

$$T_m = 81.5° \text{ C} + 16.61 \log M + 0.41 \, (\%\text{GC})$$
$$- \frac{820}{L} - 0.6 \, (\%\text{F}) - 1.4 \, (\% \text{ mismatch})$$

where M = ionic strength (mol/liter), %GC (guanosine/cytosine) refers to the mole percentage of G/C base pairs (bp) occurring in the probe, L = probe length (in bases), %F refers to the percentage of the duplex destabilizing agent formamide in the mix, and % mismatch refers to the percentage of noncomplementary base pairs between the hybridizing strands. The effect of each of these factors on hybrid stability is discussed in the following.

Increased ionic strength has a stabilizing effect on nucleic acid duplexes. This is because of neutralization of the electrostatic repulsive forces between the negatively charged phosphate groups on opposing strands. In the presence of salt, sodium anions cluster around these phosphate groups, lessening the repulsive forces they generate. Elevated salt concentrations also stabilize hybrids by decreasing the solubility of bases, thus strengthening hydrophobic interactions between them.

Hybrid stability is also affected by the percentage of GC base pairs. GC base pairs are stabilized by three hydrogen bonds, whereas AT (adenosine/thymidine) base pairs contain two. Duplexes with high GC content are more stable because of the increased number of hydrogen bonds they possess. Formamide is capable of disrupting hydrogen bonds and thus has a destabilizing effect on duplexes, reflected in a lowering of T_m. It is included in most *in situ* hybridization reactions as a means of preventing the association of nonhomologous strands at the relatively low temperatures required to maintain adequate tissue morphology.

Probe length is another important factor in determining hybrid stability. Long complementary sequences, which contain more hydrogen bonds, are more stable as a result. Synthetic oligonucleotides of 20 bases or fewer can also be used as *in situ* hybridization probes (the advantages of long over short probes for *in situ* hybridization will be discussed in a later section). The hybrids formed by these probes are less stable than the preceding equation predicts because of a decreased tendency for orderly base stacking in short duplexes. For this reason, base pair mismatches in hybrids formed by oligonucleotides are much more destabilizing than those for longer probes. Although a reduction in T_m of approximately 1 degree Celsius per percentage point base pair mismatch occurs for hybrids longer than 150 bp (Bonner et al., 1973), a reduction of 5 degrees Celsius for each mismatched base pair is observed in hybrids shorter than 20 bp (Wallace et al., 1981).

The preceding T_m equation applies to DNA–DNA duplexes. However, for the detection of RNA by *in situ* hybridization, DNA–RNA or RNA–RNA hybrids are formed when DNA or RNA probes are used, respectively. Solution hybridization studies have demonstrated that RNA-containing duplexes differ in thermal stability from DNA–DNA hybrids. RNA–RNA hybrids have been shown to be 10 to 15° C more stable than DNA–DNA duplexes of similar length and base composition (Wetmur et al., 1981; Cox et al., 1984). DNA–RNA hybrids appear to be intermediate in T_m between DNA–DNA and RNA–RNA hybrids (Casey and Davidson, 1977). Ionic strength, GC content, length, and mismatch similarly affect the thermal stability of DNA–DNA and RNA-containing duplexes. Formamide, however, appears to have a greater destabilizing effect on DNA–DNA duplexes than on others (Cox et al., 1984; Casey and Davidson, 1977).

In performing hybridization reactions, the stringency of the hybridization and wash conditions must be given careful consideration. Stringency refers to the degree to which reaction conditions favor the dissociation of nucleic acid duplexes. Stringency may be elevated by increasing temperature, decreasing salt concentration, increasing formamide concentration, and so on. A duplex formed by two nucleic acid strands with a high degree of base homology will withstand high stringency conditions better than a duplex with low homology. Hybridization reactions are generally performed at the relatively low stringency of $T_m - 25°$ C, at which point nonspecific interactions between less homologous strands may occur. A series of posthybridization washes at increasing stringencies is often employed, and the dissociation of imperfect hybrids is monitored.

Kinetics of Hybrid Formation

In most hybridization reactions, the probe is present in vast excess over the target, so that the rate of hybrid formation follows pseudo-first-order kinetics for single-stranded probes. The relationship of probe length, probe concentration, and probe complexity (number of base pairs in a nonrepeating sequence) to the time required for half the probe to hybridize to immobilized nucleic acid ($t_{1/2}$) is

$$t_{1/2} = \frac{N \ln 2}{3.5 \times 10^5} \times L^{0.5} \times C$$

where $t_{1/2}$ is expressed in seconds, N represents probe complexity, $L =$ probe strand length, and $C =$ probe concentration in mol nucleotides per liter (Meinkoth and Wahl, 1984). Absent from this equation is the target nucleotide concentration, which has no bearing on hybridization rate in the case of vast probe excess. This equation suggests that hybridization occurs most rapidly with long probes of low complexity at high concentrations. The previous relationship does not hold strictly for rates of *in situ* hybridization, which generally occurs more slowly than the equation would predict. This is most likely a result of the limited diffusion of the probe in the tissue section. For this reason, longer probes may actually result in slower rates of hybrid formation *in situ*.

The preceding considerations pertain to single-stranded probes. When the probes are double-stranded, the kinetics of hybridization become more complicated. These probes must be boiled to dissociate the strands before hybridization takes place. Two competing reactions result: (1) mixed-phase hybridization between probe and target and (2) reannealing of complementary probe strands in solution. Reannealing eliminates the probe from the hybridization reaction and slows the rate of hybrid formation. The apparent rate of hybridization with double-stranded probes may be greatly increased by the inclusion of the anionic polymer dextran sulfate to the hybridization solution. Dextran sulfate promotes "networking," the formation of probe aggregates resulting from association of partially overlapping sequences between probe molecules. This effect, which can greatly amplify hybridization signals, is thought to be caused by exclusion of DNA from the volume occupied by the polymer, thus increasing the effective concentration of the DNA (Wahl et al., 1979).

Other factors may also influence the hybridization rate. The temperature resulting in the greatest rate of hybridization has been empirically determined to be 25° C below the T_m value (Wetmur and Davidson, 1968). This is high enough to disrupt random, intrastrand hydrogen bonds within denatured DNA, but not so high as to interfere with interstrand pairing of complementary bases. Base mismatches may also affect the hybridization rate; with a twofold rate reduction for each 10 degrees Celsius in T_m resulting from sequence divergence (Bonner et al., 1973). Ionic strength has little effect on rate if it is maintained between 0.4 and 1.0 M sodium chloride (NaCl), and pH has little effect between 5.0 and 9.15 (Wetmur and Davidson, 1968).

IN SITU METHOD

The newcomer to *in situ* hybridization procedures is confronted with a bewildering array of seemingly complex and dissimilar protocols. All procedures share the goals of rendering tissue permeable to probe without the loss of RNA, of retaining good morphology, of designing probes that can effectively penetrate tissues and hybridize specifically, and of designing appropriate controls to verify the specificity of hybridization. An outline of the basic steps common to most hybridization procedures is shown in Table 1-1. It is important to note that these steps are interdependent; a change in one may require changes in subsequent steps for optimal results. Although the "ideal protocol" for a given system must be worked out empirically, an appreciation of the rationale underlying the wide variety of current approaches will speed its attainment.

Tissue Preparation

The considerations involved in tissue handling for *in situ* hybridization are critical and provide an added element of complexity relative to standard mixed-phase hybridization techniques. The preparation of tissue for *in situ* hybridization has three main goals: (1) to prevent loss of nucleic acids from the tissue, (2) to preserve tissue morphology, and (3) to allow penetration

Table 1-1   General *in situ* hybridization protocol

Tissue preparation
    Precipitating fixatives
    Cross-linking fixatives
Permeabilization
  Deproteination
    Proteases
    HCl
  Delipidation
    Detergents
    Ethanol
Probe preparation
  Probe type selection
    cDNA
    RNA
    Oligonucleotide
  Probe labeling
    Label selection
      Isotopic (^{3}H, ^{35}S, ^{32}P)
      Nonisotopic (biotin)
    Labeling method
      cDNA: nick translation, random priming
      RNA: riboprobe transcription vectors
      Oligonucleotides: T4 polynucleotide kinase, terminal deoxynucleotidyl transferase
  Length adjustment
    cDNA: DNase I
    RNA: limited alkaline hydrolysis
Background reduction treatments, e.g., acetylation
Prehybridization
Hybridization
  Stringency of hybridization conditions
  Probe concentration
Posthybridization treatments
  Enzymatic
    cDNA: S1 nuclease
    RNA: RNase
  Washes
  Dehydration
Detection
    X-ray film
    Emulsion dipping

of probe to its target. The third goal is often at odds with the other two; under certain conditions interventions that permeabilize tissues may enhance the leaching of RNA and the loss of adequate morphology. Because of the opposing nature of these factors and variations in the physical properties of different tissues, ideal tissue preparation procedures must be empirically determined.

A wide variety of methods have been used to prepare tissues for *in situ* hybridization. Perfusion-fixed tissues, fresh-frozen sections, and paraffin-embedded fixed tissues have all been employed successfully. Two types of fixatives are generally used for *in situ* hybridization: precipitating fixatives such as ethanol/acetic acid or Carnoy's and cross-linking fixatives such as paraformaldehyde, formaldehyde, and glutaraldehyde. These fixatives vary with respect to tissue permeability and nucleic acid retention. The precipitant fixatives provide best probe penetration but under some circumstances may permit the loss of RNA and of adequate tissue morphology (Lawrence and Singer, 1985). These fixatives have been used successfully for the preparation of tissue for detection of viral nucleic acid (Haase et al., 1984). The aldehyde fixatives promote cross-linking of proteins and generally provide better RNA retention and morphology than the precipitant fixatives. Glutaraldehyde provides the best RNA retention and tissue morphology, but because of extensive cross-linking, probe penetration is a problem. This must be overcome with permeabilization treatments of the type described later. The formaldehyde fixatives provide a compromise between permeability and RNA retention. Some investigators find that fixation in 3% formaldehyde provides adequate morphology and sufficient permeability to make permeabilizing treatments unnecessary (Shivers et al., 1986c). This is particularly true when small probes such as oligonucleotides are used (Lewis et al., 1985).

The treatment of tissue after fixation is one of the more variable elements of *in situ* hybridization protocols. The need to facilitate probe diffusion depends on the type and extent of fixation, the particular tissue used, the length of the probe, and section thickness. Generally, a permeabilization step is required following glutaraldehyde fixation to reduce the extent of protein cross-linking thought to create a barrier to diffusion. The need for such treatments in tissue fixed with ethanol/acetic acid or formaldehyde is more variable. These treatments may involve exposure to dilute acids, detergents, alcohols, and proteases (proteinase K, pronase, pepsin). The most commonly used methods of deproteination are treatments in dilute HC1, the enzyme proteinase K, or both. Excessive deproteination, however, may result in decreased RNA retention (Lawrence and Singer, 1985; Shivers et al., 1986c) and deterioration of morphology. The extent of tissue digestion required in an individual system must be determined empirically.

Probe Selection

Selecting a proper probe for *in situ* hybridization can be a difficult and critical decision. In recent years, an increasing variety of nucleic acid probe types have become available. Three main classes of probe are in current use: (1) complementary DNA (cDNA) probes, (2) RNA probes (riboprobes), and (3) oligonucleotide probes. Table 1-2 lists the advantages and disadvantages of each type. Riboprobes and cDNA probes are products of cloning procedures, whereas oligonucleotides may be produced by DNA synthesizers. A decision must also be made as to the type of label to attach to the probe. Radioisotopes are generally used—in particular, ^{32}P, ^{35}S, and ^{3}H, each with particular advantages and disadvantages. Procedures using nonisotopic la-

Table 1-2 *In situ* hybridization probes

Probe type	Advantages	Disadvantages
cDNA	Ease of use Networking High specific activity	Reannealing in solution Presence of vector se- quences
RNA	Single-stranded Stable hybrids High specific activity	Probe "stickiness" Familiarity with molecular biology required
Oligonucleotide	Single-stranded Easily obtained May be "tailor-made" Good tissue penetration	Access to DNA synthesizer required Dependence on published sequences Less stable hybrids

bels are currently under investigation and offer many potential advantages (Singer and Ward, 1982).

The most commonly used probes are cDNAs. The probe is double-stranded, and the "insert" (the region complementary to mRNA) is usually contained within a plasmid or bacteriophage. This type of probe has several advantages. Once obtained, these probes are relatively simple to label and use. Under certain conditions, a significant amplification of signal may be obtained with double-stranded cDNA probes. This effect may be very pronounced when long nick-translated probes in the range of 1500 bp are used (Lawrence and Singer, 1985; Meinkoth and Wahl, 1984). Nick translation (described later) produces a heterogeneous population of probe strands, many of which have overlapping complementary regions. Signal amplification may occur if a hybridizing strand has a free end capable of base pairing to a complementary region of another probe molecule. A network of interlocking strands can then accumulate about the hybridization site and amplify the signal severalfold (Lawrence and Singer, 1985).

Despite the cDNA probes' benefits, they possess some serious disadvantages. Because the strands of a duplex nucleic acid are not free to hybridize, cDNA must be denatured by boiling before it is applied to tissue sections. Only half the strands—those that are complementary to the mRNA—are able to hybridize. The presence of both strands in the hybridization mix results in two competing reactions: mixed-phase hybridization between probe and target, and probe reannealing in solution. Cox et al. (1984) have demonstrated that probe reannealing may be the favored reaction, so that only a small percentage of probe is available to participate in hybridization. The presence of vector DNA is another drawback to the use of cDNA probes. Labeled vector sequences may become nonspecifically trapped in tissue sections and increase background levels of staining. If high backgrounds are a problem, however, the cloning vector may be removed from the insert.

The use of RNA probes for *in situ* hybridization has been pioneered largely by Angerer and colleagues (see Cox et al., 1984). Riboprobes are single-stranded RNA molecules produced from a cloned cDNA that has been introduced into a specially designed plasmid transcription system (described later). Because these probes are single-stranded, reannealing in solution does not occur, so a greater percentage of probe is available for hybridization. Perhaps for this reason, riboprobes may produce stronger signals than cDNA probes in some systems (Cox et al., 1984). In addition, the RNA–RNA hybrids produced are more stable than corresponding DNA–RNA hybrids (Wetmur et al., 1981). The transcription reaction produces labeled probe of a fixed length and high specific activity. These probes are "stickier" than DNA probes and produce a higher degree of nonspecific binding to tissue; therefore, posthybridization RNase treatment is required to reduce background. A potential drawback to using riboprobes is that their production requires greater facility with molecular biology procedures than is required for other types of probes.

Because of the recent advent of automated DNA synthesizers, short (generally 15–50 base pairs), single-stranded DNA oligonucleotides have become available for *in situ* hybridization (Coghlan et al., 1985; Lewis et al., 1985). These probes are quickly and easily made, compared with the substantial effort normally involved in the cloning procedures required to obtain cDNA and riboprobes. Their short length makes oligonucleotides better able to penetrate tissue sections. The ability to tailor an oligonucleotide to precisely the portion of an mRNA desired allows for probes of great specificity, which can be useful in discriminating an mRNA from other members of its gene family. Potential drawbacks to the use of oligonucleotide probes include availability of a DNA synthesizer, the possibility of errors in published mRNA sequences, and the need to establish less stringent hybridization and wash conditions for short probes which may result in less specificity.

Probe Labeling

Of the many strategies developed for the labeling of probes, most employ radioisotopes (nonisotopic labeling with biotin is discussed in a later section). The choice of a radioisotope is guided by requirements for sensitivity, anatomical resolution, and speed of development.

^{32}P-labeled probes may achieve high specific activities and are the most energetic probes in current use. Relatively short autoradiographic exposure times (1–7 days) are sufficient with these probes. A problem with ^{32}P-labeled probes is that the long path length of the energetic beta emissions results in anatomical resolution that is generally inadequate for localization at the single cell level (see Coghlan et al., 1985, for a notable exception). Another drawback to using these probes is their short half-life (2 weeks). Such probes are excellent however, for regional localization using x-ray film.

^{3}H-labeled probes are commonly used for *in situ* hybridization. They offer excellent anatomical resolution and low backgrounds. However, their low specific activities and low-energy emissions require long exposure times (weeks to months) to detect targets low in number.

Probes labeled with ^{35}S may attain higher specific activities and greater signals than ^{3}H-labeled probes and permit cellular localization. However, some investigators have noted higher levels of nonspecific binding with these probes.

Nick translation is the most commonly used method for labeling cDNA probes (Rigby et al., 1977). The method depends on the activities of the enzymes pancreatic DNase I and *E. coli* polymerase I. DNase I makes cuts in one strand of a duplex (nicks) at random sites throughout the DNA molecule. The DNA polymerase then mediates the synthesis of DNA through two actions: a $5' \rightarrow 3'$ exonuclease activity and a $5' \rightarrow 3'$ polymerase activity. By these two actions, radioactive DNA is synthesized through the incorporation of [3]H-, [32]P-, or ^{35}S-labeled nucleotides. Uniform labeling along the length of the molecule results. Because of the random nature of the nicking, denaturation produces single strands of varied lengths from random portions of the molecule. The average strand length is determined by the concentration of DNase I. Another similar technique, involving the random priming of single-stranded DNAs, has recently been developed and provides probes of higher specific activity (Feinberg and Vogelstein, 1983).

RNA probes may be synthesized and labeled in reactions involving the use of specially designed transcription vectors that contain prokaryotic RNA polymerase promoters. Two such vectors are the plasmids pSP64 and pSP65, which contain a multiple cloning site, a short nucleotide sequence containing the recognition sequences for a variety of restriction endonucleases. This site allows for the precise insertion (subcloning) of the cDNA of interest into the plasmid. To the $5'$ side of the insert is the recognition sequence (promoter) for the RNA polymerase enzyme of the *Salmonella* SP6 phage. For transcription, the circular plasmid is linearized by a restriction endonuclease digestion at the $3'$ end of the insert or at a chosen site within the insert. Under appropriate reaction conditions, the SP6 RNA polymerase enzyme will recognize its promoter and transcribe in the $5' \rightarrow 3'$ direction until it reaches the site of linearization. The enzyme will incorporate free ribonucleotides into a single-stranded RNA transcript of the cDNA insert. High specific activity probes may be produced when radiolabeled (^{3}H, ^{32}P, or ^{35}S) ribonucleotides are included in the reaction mix. The resultant probes are uniformly labeled and of a fixed length.

The plasmids pSP64 and pSP65 differ only in the orientation of the multiple cloning site in relation to the SP6 promoter. Subcloning a cDNA insert into both plasmids enables either strand of the insert to be transcribed separately. The "antisense" strand, complementary to cellular mRNA, is an effective hybridization probe (DeLeon et al., 1983). A "sense" strand may also be produced that is similar in length, specific activity, and base composition, thus providing an excellent negative control. Recently introduced transcription vectors (e.g., Gemini vectors, ProMega Biotech) possess two different promoters located on either side of the multiple cloning site. This allows for the transcription of sense and antisense probes from the same plasmid. This vector carries promoters for the SP6 and T7 RNA polymerase enzymes on opposite ends of its multiple cloning site. These promoters are oriented so that SP6 polymerase will transcribe a strand of the insert in one direction whereas T7 polymerase will transcribe in the opposite direction, copying the opposite strand of the insert. If necessary, the length of the transcripts may subsequently be reduced by limited alkaline hydrolysis.

Oligonucleotides may be ^{32}P-labeled to high specific activities through the use of T4 polynucleotide kinase (Maniatis et al., 1982). This enzyme catalyzes the transfer of the gamma-phosphate of adenosine triphosphate (ATP) to the 5′-hydroxyl terminus of the oligonucleotide. Another method for labeling oligonucleotides for *in situ* hybridization uses the enzyme terminal deoxynucleotidyl transferase (Bollum, 1974). This enzyme may catalyze the addition of ^{3}H-, ^{32}P-, or ^{35}S-labeled deoxynucleotides to the 3′-end of oligonucleotides (Lewis et al., 1985). A probe may achieve high specific activity by adding a polymer of labeled bases. The presence of these noncomplementary tails does not appear to have a significant adverse effect on hybridization *in situ* (Lewis et al., 1985).

Treatments to Decrease Nonspecific Binding

Critical to the sensitivity of *in situ* hybridization techniques is the minimization of background. Unless signal-to-noise ratios are favorable, it is difficult to discern low-level signals, which may require 1 to 2 month exposure times for detection. Background can arise from a variety of sources: from the formation of imperfect duplexes with nonhomologous nucleic acids, from electrostatic interactions among charged groups, from physical entrapment in the three-dimensional lattice of the tissue section, and from artifacts of a detection system. The nonspecific retention of probe in tissue sections may be affected by at least four factors: (1) treatments to decrease tissue "stickiness"; (2) treatments to block nonspecific binding of probe; (3) posthybridization enzyme treatments; and (4) posthybridization washes. The last two factors are discussed in later sections.

The "stickiness" of a tissue may depend in part on electrostatic attraction between the hybridization probe and basic proteins in the tissue section. This possible source of background is minimized by some investigators by treatment with 0.25% acetic anhydride. Theoretically, acetic anhydride blocks basic groups by acetylation (Hayashi et al., 1978). Lawrence and Singer (1985) have found acetic anhydride treatment of tissue sections to greatly reduce backgrounds for probes of more than 1500 nucleotides, but not for smaller probes. Some investigators also acetylate slides and coverslips to minimize nonspecific adherence of probes to glass (Haase et al., 1984).

Most *in situ* hybridization protocols call for a prehybridization step designed to decrease background binding of the probe. The components of the prehybridization mix are intended to saturate sites in the tissue section that might otherwise bind nucleic acids nonspecifically. Prehybridization solutions are similar to those used in filter hybridizations and effectively decrease nonspecific binding to glass slides and tissue sections (Brahic and Haase, 1978). They typically include ficoll, bovine serum albumin, and polyvinyl pyrrolidone to decrease nonspecific binding to proteins, polysaccharides, and nucleic acids. Sodium pyrophosphate, nucleic acids, and ethylenediaminetetraacetic acid, (EDTA) are also added to decrease nonspecific nucleic acid interactions. Prehybridization mixes are usually boiled and quick-cooled before they are applied to denature their nucleic acid components. The prehybridization step is commonly carried out at the same temperature as the hybridization step, for a duration of 2 to several hours.

Hybridization Considerations

The goal of most of the previous steps has been to provide optimal conditions for the hybridization reaction. Many factors must be taken into account in planning this step. The principles of solution hybridization are important, as well as considerations unique to *in situ* hybridization. Other important factors include probe concentration, probe length, stringency of hybridization conditions, kinetics of *in situ* hybridization, and preservation of tissue morphology.

Hybridization mixes generally include all the components of prehybridization mixes, as well as labeled probe and dextran sulfate. When double-stranded cDNA probes are used, they must be denatured by boiling to separate the strands before hybridization. This step is not necessary for single-stranded RNA or for oligonucleotide probes. Dextran sulfate increases the rate and signal amplification of hybridization by an excluded volume effect, as discussed previously.

The stability of hybrids formed *in situ* exhibits the same properties that characterize liquid phase and filter hybridization (Brahic and Haase, 1978). However, a slight depression of 5° from the calculated T_m of DNA–RNA hybrids (Brahic and Haase, 1978) and RNA–RNA hybrids (Cox et al., 1984) has been noted for duplexes formed *in situ*. It has been hypothesized (Cox et al., 1984) that this depression results from cross-linking of mRNA molecules in the fixed tissue section. Such cross-links could form obstructions that prevent the full length of the probe from hybridizing, resulting in the formation of shorter-than-expected duplexes. As discussed previously, duplex length is directly related to T_m.

Like filter hybridization, *in situ* hybridization has been found to occur under low stringency conditions, at an optimal temperature of $T_m - 25°$. Because of their increased stability, RNA–RNA hybrids require higher hybridization temperatures (by about 10–15°) than DNA–RNA hybrids of comparable length and GC content (Cox et al., 1984). It also follows that oligonucleotide probes require lower hybridization temperatures than cDNA or riboprobes. Most hybridization mixes contain 50% formamide, so that optimal stringencies may be obtained without the need to elevate temperatures greatly. Prolonged exposure to high temperatures causes a deterioration of cell morphology as well as loss of adherence of the section to the slide. Hybridization is generally performed in the dark because ionization of formamide is enhanced by light.

The kinetics of *in situ* hybridization and of filter hybridization differ significantly. Using cDNA probes, Brahic and Haase (1978) found the rate of hybridization *in situ* to be an order of magnitude slower than that of liquid-phase hybridization. This difference has been attributed to decreased access to the target nucleic acid in the tissue section. Cox et al. (1984) compared the signals resulting from hybridization with cDNA and riboprobes and found eightfold greater signals with the single-stranded RNA probes. They posited that reannealing of the cDNA in solution quickly removed most of the probe from the hybridization reaction. The kinetics of riboprobe hybridization *in situ* was also found to be unusual. Maximum signal was seen to be proportional to probe concentration even at "saturating" probe concentrations, indicating that only a small fraction of single-stranded probe was free to hybridize.

The optimal probe concentration for *in situ* hybridization is that which produces the greatest signal-to-noise ratio. Because background may be linearly related to probe concentration (Cox et al., 1984), the desired probe concentration is therefore the lowest required to saturate target nucleic acid. Less probe would result in a decreased signal, and more would only increase the background. Because most of a hybridization probe is not free to hybridize, the optimal concentration is difficult to predict and must be determined empirically.

Several considerations enter into the selection of probe length. In general, with standard tissue fixation and permeabilization methods, probes less than 300 to 400 nucleotides in length may adequately penetrate tissue. Lawrence and Singer (1985) found little effect of cDNA probe length on the signal except for probes longer than 1500 nucleotides. Probes of this size produced an amplification of signal through the networking phenomenon discussed previously. This is a desirable effect when maximal sensitivity is sought. Oligonucleotides provide maximal tissue penetration but may produce weaker signals than are possible with uniformly labeled long hybridization probes.

Posthybridization Treatments

As with filter hybridization, the hybridization step *in situ* is performed under low-stringency conditions. These conditions may permit the nonspecific adherence of probe molecules to various tissue components. The aim of posthybridization treatments is to reduce the background through wash steps, and in some cases through treatments with nucleases that specifically degrade single-stranded nucleic acid. RNA probes in particular tend to exhibit high levels of nonspecific binding, which are most effectively reduced by posthybridization treatment with RNase (John et al., 1969; Lynn et al., 1983). RNase, used under appropriate conditions, selectively removes non-base-paired RNA from tissue sections, so that backgrounds may be greatly reduced with little loss of signal. Posthybridization enzyme treatments are not commonly performed when double-stranded cDNA probes are used; however, the enzyme S1 nuclease, which acts on single-stranded nucleic acids, has been effectively employed (Godard, 1983).

Posthybridization washing conditions should be chosen to minimize background without reducing the hybridization signal. Optimal wash conditions are determined by numerous factors, including method of fixation, probe concentration, probe length, and hybrid stability. The acceptable level of background may differ depending on the length of the intended exposure; if a long exposure is planned, more extensive washing may be required. Washes should attain greater stringencies than were used for hybridization to reduce nonspecific adherence to tissue components and to promote the dissociation of weakly complementary hybrids. A series of washes of increasing stringency is typically performed by varying salt concentration, temperature, or formamide concentration. The stringency required for optimal signal-to-noise ratio must be determined empirically. Following the wash steps, sections are generally dehydrated, dipped in emulsion, or apposed to x-ray film.

Determination of Hybridization Specificity

It is essential to verify the specificity of hybridization signals by performing appropriate controls. Specific hybridization may be easily confused with unwanted hybrid formation between the probe and weakly homologous sequences or with nonspecific interactions between probe and nonnucleic acid tissue components. A number of criteria may be used to assess the specificity of hybridization: pretreatment of the tissue with RNase, use of heterologous (unrelated) probes, use of multiple probes for a single target, melting-point analysis of hybrids formed *in situ*, or correlation with immunohistochemistry. A combination of these strategies will provide a strong indication of specificity.

At times, nonspecific sticking of probe to tissue components other than RNA may be difficult to distinguish from a true hybridization signal. Pretreatment of tissue sections with RNase should eliminate true hybridization signals, with little effect on nonspecific sticking. However, this control does not rule out nonspecific binding to nontarget nucleic acids. The use of a heterologous probe may help to verify that a hybridization signal is the result of specific base pairing between probe and target. Such a probe should be similar in length, GC content, and specific activity to the probe under study. The lack of a signal with such a control verifies that probe binding is a result of its base sequence and not its physical properties. Ideal probes for this purpose are riboprobe sense and nonsense strands. One should be aware that accidental sequence homologies between heterologous probes and cellular RNAs sometimes occur.

In some settings, specially designed probes complementary to nonoverlapping regions of the target nucleic acid sequence may provide a positive control. These probes, which are capable of binding to separate regions of the same target mRNA, may then be hybridized to alternate tissue sections. If, on adjacent sections, the same individual cells hybridize with these different probes, specificity is further confirmed.

A valuable demonstration of hybrid specificity can be obtained through competition between labeled probe and an excess of the unlabeled sequence. Theoretically, nonspecific binding sites in tissue are present in large numbers, whereas only a much smaller, fixed number of target sites are available for probe binding. Excess unlabeled probe would be expected to compete effectively with labeled probe for specific hybridization sites, without having a significant effect on the nonspecific sticking of probe molecules. A decrease in signal in a competed section would thus be attributable to the blocking of specific hybridization by the competitor. If competition does not affect the signal, then the specificity of the signal must be questioned.

Further verification of *in situ* hybridization specificity may be obtained by analyzing hybrid thermal stability. As discussed, the T_m of a hybrid formed *in situ* should conform closely to the value predicted for liquid hybridization. Because the desired hybrids are better base paired than others, they should resist more stringent environments. Melting curve analysis (plotting the percentage of duplex molecules as a function of temperature) should provide a T_m close to the predicted value. A decrease in signal at lower-than-expected temperatures reflects the dissociation of heterologous hybrids. Cox et al. (1984) demonstrated the degree of precision this type of analysis permits; in one case a clear 12° decrease in T_m for a probe with a 10 to 15% sequence divergence relative to a homologous probe was determined.

Finally, evidence of hybridization specificity can also result from a correlation between immunocytochemical and *in situ* hybridization data on a cell-by-cell basis. This has been accomplished on alternate thin sections for proopiomelanocortin (POMC) mRNA and gene products (Gee and Roberts, 1983), as well as in the same section for viral antigens and RNA (Brahic et al., 1984). Simultaneous detection of peptide hormone antigens and mRNA has also been demonstrated (Wolfson et al., 1985; Sherman et al., 1984; Shivers et al., 1986b). Immunocytochemical and *in situ* hybridization results, however, will not always be correlated. As mentioned previously, there are instances in which cells containing a particular mRNA fail to exhibit staining for the protein for which it codes. Conversely, cells may concentrate a particular protein or peptide and thus evidence immunostaining without an *in situ* hybridization signal (Gee et al., 1983).

FUTURE DIRECTIONS

Because of the current rapid progress in the field of molecular cloning, *in situ* hybridization probes are becoming available for the study of a large number of biologically important proteins. *In situ* hybridization methods are also being refined at a rapid pace, so it may be expected that this increasingly powerful technique will soon be put to widespread use. Future methodological advances that are likely to occur include improvements in the effective utilization of nonisotopic probes and the development of procedures that provide truly quantitative data. Such methods will be used to study regulation of gene expression during development, transcriptional responses to neural activity, hormonal or pharmacological stimuli, chromosomal mapping in cytological preparations, and the molecular biology of a wide variety of disease states.

While *in situ* hybridization techniques are currently capable of detecting qualitative changes in nucleic acid concentration, grain counts cannot be well correlated with absolute levels of nucleic acids. Part of the difficulty lies in the multistep nature of the method; for the results to be reproducible, one must avoid small variations in performing a variety of steps. Many factors must be carefully considered and controlled if quantitative data are to be collected: section thickness, nucleic acid retention, consistency of hybridization, exposure and development conditions, and the thickness and uniformity of emulsion coating. Essential for the acquisition of quantitative data is the use of standard curves, which permit the correlation of the grain density resulting from a particular procedure with the absolute levels of nucleic acids.

Another promising development in *in situ* hybridization methodology is the use of nonisotopic probes. The most commonly used nonisotopic label is biotin (vitamin H of the B_2 complex), which will bind to the glycoprotein avidin with very high affinity. This strong specific interaction has been exploited using avidin to localize biotinylated nucleic acids *in situ* (Singer and Ward, 1982; Langer-Safer et al., 1982). The approach offers several potential advantages, including speed, safety, and high resolution. The detection procedure normally takes one day, in contrast with the prolonged autoradiographic exposures (often several months) required when ^{3}H-labeled probes are used. In addition, one can avoid the hazards of working with radioisotopes with this method and achieve a higher degree of spatial resolution.

Biotin labeling may be accomplished with the use of a biotinyl-d uridine 5'-triphosphate (dUTP) derivative containing an 11- or 16-carbon spacer arm between the pyrimidine ring and the biotin moiety. The spacer increases the availability of the biotin to the avidin or antibiotin antibody detection molecules (Singer and Ward, 1982). Incorporation of biotinyl-dUTP is typically achieved by nick translation (Langer et al., 1981). A recently introduced method for the biotinylation of nucleic acids uses photobiotin, a photoactivatable analog of biotin (Forster et al., 1985). Photobiotin forms stable linkages with nucleic acids upon irradiation with visible light. This reaction is simply performed, brief, applicable to single- or double-stranded DNA or RNA, and inexpensive in terms of enzymes and substrates.

The *in situ* hybridization of biotinylated probes can be detected by a variety of methods. Avidin has been conjugated to rhodamine for fluorescent detection of hybrids and to biotinylated peroxidase for enzymatic detection.

Hybrids may also be detected by an antibiotin antibody, followed by a peroxidase or rhodamine-conjugated second antibody (Singer and Ward, 1982; Langer-Safer et al., 1982). Biotinylated hybrids may also be detected with an electron microscope using a peroxidase-conjugated second antibody (Manuelidis et al., 1982) or using a second antibody conjugated to colloidal gold (Hutchison et al., 1982).

A major barrier to the widespread use of these methods is a lack of sensitivity compared to that of standard autoradiographic techniques. Although biotinylated probes may detect nucleic acids with a high degree of sensitivity on nitrocellulose blots (Leary et al., 1983), this is not generally achievable *in situ*. The development of detection methods that greatly amplify signals will thus be of great benefit. An approach we are taking to accomplish this goal is the adaptation of immunogold-silver methods (Danscher, 1981) for *in situ* hybridization. This technique has been demonstrated to provide an order of magnitude increase in sensitivity over standard immunocytochemical methods (Hacker et al., 1985). We are applying this method to *in situ* hybridization, detecting biotinylated probes by employing silver enhancement of an antibiotin–colloidal gold complex.

A problem to which new *in situ* hybridization techniques may be successfully applied is the regulation of gene expression. As mentioned earlier, regulation of protein production and release may be correlated with alterations in mRNA levels. However, changes in mRNA levels may take several days to occur, long after the initial effects of a regulator are detectable. Changes in the levels of heterogeneous nuclear RNA (hnRNA: primary gene transcripts subject to extensive splicing within the nucleus) may occur within a few hours of exposure to a regulatory stimulus and may provide the earliest indication of a cellular response (Eberwine and Roberts, 1984). Probes complementary to intron sequences (those regions of hnRNA that are removed in the nucleus) may enable one to monitor hnRNA levels by *in situ* hybridization. Because of the rapid pace of molecular cloning, probes now becoming available will enable the study of the regulation of a variety of components of a functional system. For example, new insights into the coordinated action of a neuropeptide system may be obtained by *in situ* hybridization with probes for the hormone, its processing enzymes, and its receptor.

In situ hybridization will be increasingly useful to medical researchers as a tool for the investigation of the pathophysiology of many disease states. It will aid the characterization of disease processes that are clearly of viral origin and the study of chronic diseases in which viruses have been implicated. Examples of this approach include the demonstration of measles virus sequence in the brains of multiple sclerosis patients (Haase et al., 1981) and the discovery of HTLVIII viral sequences in the lymphocytes of victims of acquired immune deficiency syndrome (AIDS) (Harper et al., 1986).

The methodology may also prove quite important for the study of various forms of neurological and mental disorders. In these instances, *in situ* hybridization may be a powerful research tool to determine changes in protein and peptide systems in terms of gene expression in highly specific cell groupings. Preliminary information regarding the stability of messenger RNA in autopsy material is encouraging, although the topic requires considerable additional investigation. As the range of probes, sensitivity, and quantitation procedures are further developed, the method can be expected to permit one to gather information that could not be obtained in any other way. *In situ* hybridization may be particularly valuable for studies dealing with efforts to establish subtypes of various disorders. Determining the subtypes of severe mental disorders, such as the putative multiple forms of affective disorders, psychoses, and severe disorders of children, particularly highlights such potential applications. Thus, in the future we see the possibility that the method will permit determination of alterations in specific gene expression in highly localized areas of brain in association with particular forms of neurologic and mental disease.

In addition, *in situ* hybridization may become a useful diagnostic tool. It could have utility for the diagnosis of viral infections in cases in which no immunological responses are apparent. *In situ* hybridization performed with cytological preparations may one day aid in the diagnosis of genetic diseases. With continued improvements in the ease and rapidity with which the technique may be performed, *in situ* hybridization will become a routine procedure in the clinical pathology laboratory.

As the technique of *in situ* hybridization becomes increasingly quantifiable, it has power to impact on the development of behavioral neurochemistry. The biochemical and anatomical specificity of the technique and its ability to permit exploration of gene expression make it particularly suited to studies of the ways in which behavioral events change the brain. Using this technique, very subtle changes may be examined in areas so small that they could not otherwise be studied chemically. In this sense, the method may permit consideration of the sequential changes in gene expression related to behavior. *In situ* hybridization will facilitate determination of the pathways used and recognition of the neuroregulators, or other proteins and peptides, which are altered by a highly defined behavioral sequence. In this sense, *in situ* hybridization will be a particularly effective arrow in the quiver of the biochemical analysis of behavioral processes.

ACKNOWLEDGMENTS

The authors wish to acknowledge Alicia Fritchle for preparing the manuscript, Mark Schaefer for his critical reading of the manuscript, and the MacArthur Foundation NIMH, NIDA, and ONR for their support.

REFERENCES

Bollum, F.J. Terminal deoxynucleotidyl transferase. In *The Enzymes*, Vol. 10, P.D. Boyer (Ed.). New York: Academic Press, 1974.

Bonner, T.I., Brenner, D.J., Neufeld, B.R., and Britten, R.J. Reduction in the rate of DNA reassociation by sequence divergence. *J. Mol. Biol.* 81:123–125, 1973.

Brahic, M. and Haase, A.T. Detection of viral sequences of low reiteration frequency by *in situ* hybridization. *PNAS* 75:6125, 1978.

Brahic, M., Haase, A.T., and Cash, E. Simultaneous *in situ* detection of viral RNA and antigens. *PNAS* 81:5445–5448, 1984.

Buongiorno-Nardelli, S. and Amaldi, F. Autoradiographic detection of molecular hybrids between rRNA and DNA in tissue sections. *Nature* 225:946–948, 1970.

Casey, J. and Davidson, N. Rates of formation and thermal stabilities of RNA:RNA and DNA:DNA duplexes at high concentrations of formamide. *Nucl. Acids Res.* 4:1539–1552, 1977.

Chaw, Y.F.M., Crane, L.E., Lange, P., and Shapiro, R. Isolation and identification of cross-links from formaldehyde-treated nucleic acids. *Biochemistry* 19:5525–5531, 1980.

Coghlan, J.P., Aldred, P., Haralambidis, J., Niall, H.D., Penschow, J.D., and Tregear, G.W. Hybridization histochemistry. *Anal. Biochem.* 149(1):1–28, 1985.

Cox, K.H., DeLeon, D.V., Angerer, L.M., and Angerer, R.C. Detection of mRNAs in sea urchin embryos by *in situ* hybridization using asymmetric RNA probes. *Dev. Biol.* 101:485–502, 1984.

Danscher, G. Localization of gold in biological tissue. *Biochemistry* 71:81–88, 1981.

DeLeon, D.V., Cox, K.H., Angerer, L.M., and Angerer, R.C. Most early-variant histone mRNA is contained in the pronucleus of sea urchin eggs. *Dev. Biol.* 100:197–206, 1983.

Eberwine, J.H. and Roberts, J.L. Glucocorticoid regulation of pro-opiomelanocortin gene transcription in the rat pituitary, *J. Biol. Chem.* 259:2166–2170, 1984.

Feinberg, A.P. and Vogelstein, B. A technique for radiolabelling DNA restriction endonuclease fragments to high specific activity. *Anal. Biochem.* 132:6–13, 1983.

Forster, A.C., McInnes, J.L., Skingle, D.C., and Symons, R.H. Non-radioactive hybridization probes prepared by the chemical labeling of DNA and RNA with a novel reagent, photobiotin. *Nucl. Acids. Res.* 13:745–761, 1985.

Fostel, J., Narayanswami, S., Hamkalo, B., Clarkson, S.G., and Pardue, M.L. Chromosomal location of a major tRNA gene cluster of *Xenopus leavis.* *Chromosoma* 90:254–260, 1984.

Gall, J.G. and Pardue, M. Formation and detection of RNA–DNA hybrid molecules in cytological preparations. *Proc. Natl. Acad. Sci. USA* 63:378–383, 1969.

Gee, C.E., Chen, C.L., Roberts, J.L., Thompson, R., and Watson, S.J. Identification of proopiomelanocortin neurons in rat hypothalamus by *in situ* cDNA-mRNA hybridization. *Nature* 306:374–376, 1983.

Gee, C.E. and Roberts, J.L. Laboratory methods for *in situ* hybridization histochemistry: A technique for the study of gene expression in single cells. *DNA* 2:155–161, 1983.

Godard, C.M. Improved method for detection of cellular transcripts by *in situ* hybridization: Detection of virus-specific RNA in Rous sarcoma virus-infected cells by *in situ* hybridization to cDNA. *Histochemistry* 77:123–131, 1983.

Haase, A.T., Brahic, M., Stowring, L., and Blum, H. Detection of viral nucleic acids by *in situ* hybridization. In *Methods in Virology*, Vol. VII, K. Maramorosch and H. Koprowski (Eds.). New York: Academic Press, 1984, pp. 189–226.

Haase, A.T., Venture, P., Gibbs, C., and Tourtellotte, W.: Measles virus nucleotide sequences: Detection by hybridization *in situ*. *Science* 212:672, 1981.

Hahn, W.E., Van Ness, J., and Chaudhari, N. Overview of the molecular genetics of mouse brain. In *Molecular Genetic Neuroscience*. F.O. Schmitt, S.J. Bird, and F.E. Bloom (Eds.). New York: Raven Press, 1982, pp. 323–334.

Hacker, G.W., Springall, D.R., Van Noorden, S., Bishop, A.E., Grimelius, L., and Polak, J. The immunogold-silver staining method: A powerful tool in histopathology. *Virchows Archiv. A* 406:449–461, 1985.

Harper, M.E., Marselle, L.M., Gallo, R.C., and Wong-Staal, F. Detection of lymphocytes expressing human T-lymphotropic virus type III in lymph nodes and peripheral blood from infected individuals by *in situ* hybridization. *PNAS* 83:772–776, 1986.

Hayashi, S., Gillam, I.C., Delaney, A.D., and Tener, G.M. Acetylation of chromosome squashes of *Drosophila Melanogaster* decreases the background in autoradiographs from hybridization with [^{125}I]-labeled RNA. *J. Histochem. Cytochem.* 26:677–679, 1978.

Hutchison, N.J., Langer-Safer, P.R., Ward, D.C., and Hamkalo, B.A. *In situ* hybridization at the electron microscope level: Hybrid detection by autoradiography and colloidal gold. *J. Cell Biol.* 95:609–618, 1982.

John, H.A., Birnstiel, M.L., and Jones, K.W. RNA–DNA hybrids at the cytological level. *Nature* 223:582–587, 1969.

Jones, K.W. The method of *in situ* hybridization. *New Techniques in Biophysics and Cell Biology* 1:29–66, 1974.

Langer, P.R., Waldrop, A.A., and Ward, D.C. Enzymatic synthesis of biotin-labeled polynucleotides: Novel nucleic acid affinity probes. *PNAS* 78:6633–6637, 1981.

Langer-Safer, P.R., Levine, M., and Ward, D.C. Immunological method for mapping genes on *Drosophila* polytene chromosomes. *PNAS* 79:4381–4385, 1982.

Lawrence, J.B., and Singer, R.H. Quantitative analysis of *in situ* hybridization methods for the detection of actin gene expression. *Nuc. Acids Res.* 15:1777–1799, 1985.

Leary, J.J., Brigati, D.J., and Ward, D.C. Rapid and sensitive colorimetric method for visualizing biotin-labeled DNA probes hybridized to DNA or RNA immobilized on nitrocellulose: Bio-blots. *PNAS* 80:4045–4049, 1983.

Lewis, M.E., Sherman, T.G., and Watson, S.J. *In situ* hybridization histochemistry with synthetic oligonucleotides: Strategies and methods. *Peptides* 6 (Suppl. 2); 75–87, 1985.

Lynn, D.A., Angerer, L.M., Bruskin, A.M., Klein, W.H., and Angerer, R.C. Localization of a family of mRNAs in a single cell type and its precursors in sea urchin embryos. *PNAS* 80:2656–2660, 1983.

Maniatis, T., Fritsch, E.F., and Sambrook, J. *Molecular Cloning*. New York: Cold Spring Harbor Laboratory, 1982.

Manuelidis, L., Langer-Safer, P.R., and Ward, D.C. High-resolution mapping of satellite DNA using biotin-labeled DNA probes. *J. Cell Biol.* 95:619–625, 1982.

Meinkoth, J. and Wahl, G. Hybridization of nucleic acids immobilized on solid supports. *Anal. Biochem.* 138:267–284, 1984.

Pardue, M.L. and Dawid, I.B. Chromosomal locations of two DNA segments that flank ribosomal insertion-like sequences in *Drosophila*: Flanking sequences are mobile elements. *Chromosoma* 83:29–43, 1981.

Rigby, P.W.J., Dieckmann, M., Rhodes, C., and Berg, P. Labeling deoxyribonucleic acid to high specific activity *in vitro* by nick translation with DNA polymerase I. *J. Mol. Biol.* 113:237–251, 1977.

Sherman, T.G., Watson, S.J., Herbert, E., and Akil, H.: The co-expression of dynorphin and vasopressin: An *in situ* hybridization and dot blot analysis of mRNAs during stimulation. *Abstracts, Soc. Neurosci.* 10:359, 1984.

Shivers, B.D., Harlan, R.E., Hejtmancik, J.F., Conn, P.M., and Pfaff, D.W. Localization of cells containing LHRH-like mRNA in rat forebrain using *in situ* hybridization. *Endocrinology* 118:883–885, 1986a.

Shivers, B.D., Harlan, R.E., Pfaff, D.W., and Schacter, B.S. Combination of immunocytochemistry and *in situ* hybridization in the same tissue section of rat pituitary. *J. Histochem. Cytochem.* 34:39–43, 1986b.

Shivers, B.D., Schachter, B.S., and Pfaff, D.W. *In situ* hybridization for the study of gene expression in the brain. In *Methods in Enzymology* 124:497–510, 1986c.

Singer, R.H. and Ward, D.C. Actin gene expression visualized in chicken muscle culture by using *in situ* hybridization with a biotinated nucleotide analog. *PNAS* 79:7331–7335, 1982.

Tank, A.W., Lewis, E.J., Chikaraishi, D.M., and Weiner, N. Elevation of RNA coding for tyrosine hydroxylase in rat adrenal gland by reserpine treatment and exposure to cold. *J. Neurochem.* 45:1030–1033, 1985.

Thomas, C.A. and Dancis, B.M. Ring stability. *J. Mol. Biol.* 77:44–55, 1973.

Uhl, G.R., Zingg, H.H., and Habener, J.F. Vasopressin mRNA *in situ* hybridization: Localization and regulation studied with oligonucleotide cDNA probes in normal and Brattleboro rat hypothalamus. *PNAS* 82:5555–5559, 1985.

Wahl, G.M., Stern, M., and Stark, G.R. Efficient transfer of large DNA fragments from agarose gels to diazobenzyloxymethyl-paper and rapid hybridization by using dextran sulfate. *Proc. Natl. Acad. Sci. USA* 76:3683–3687, 1979.

Wallace, R.B., Johnson, M.J., Hirose, T., Miyake, T., Kawashima, E.H., and Itakura, K. The use of synthetic oligonucleotides as hybridization probes. II. Hybridization of oligonucleotides of mixed sequence to rabbit β-globin DNA. *Nucl. Acids Res.* 9:879–894, 1981.

Wetmur, J.G. and Davidson, N. Kinetics of renaturation of DNA. *J. Mol. Biol.* 31:349–370, 1968.

Wetmur, J.G., Ruyechan, W.T., and Douthart, R.J. Denaturation and renaturation of *Penicillium chrysogenum* mycophage double-stranded ribonucleic acid in tetraalkylammonium salt solutions. *Biochemistry* 20:2999–3002, 1981.

Wolfson, B., Manning, R.W., David, L.G., Arentzen, R., and Baldino, F. Colocalization of corticotropin releasing factor and vasopressin mRNA in neurones after adrenalectomy. *Nature* 315:59–61, 1985.

2

Location of Gene Expression in Mammalian Cells

J. P. Coghlan, J. D. Penschow, J. R. Fraser,
P. Aldred, J. Haralambidis, and G. W. Tregear

In situ tissue hybridization techniques were introduced in 1969. To say that this was simply a logical extension of standard procedures used in molecular biology is not to denigrate the percipience of the pioneers in this field (Gall and Pardue, 1969; John et al., 1969; Buongiorno-Nardelli and Amaldi, 1970); many ideas seem obvious and simple once they are put forward. In this case the basic idea fell on fairly fallow ground, and a few papers developed. For a full blossoming to occur, widespread acceptance of molecular biological techniques was needed. The history and development of the field, including the authors' own entry, with the first pubished use of cDNA probes and the first uses of oligonucleotide probes, have been detailed elsewhere (Coghlan et al., 1984; Hudson et al., 1981; Coghlan et al., 1985; Rall et al., 1985; Jacobs et al., 1983).

We have chosen to use the term "hybridization histochemistry" to highlight the technique's similarity and complementarity to immunohistochemistry. Furthermore, readers who have used key word–based literature surveys will be aware, on the one hand, that their printouts get diluted by other, nontissue *in situ* hybridization procedures, and on the other hand, that many relevant papers are lost because they were entered under less obvious or less satisfactory bywords.

Many of the problems that originally confronted the pioneers of the technique have since been solved. For instance, cDNA and oligonucleotide probes are now readily available as well as a variety of isotopic and nonisotopic labels, and some improvements have been made in histological procedures. Technical improvements have brought significant benefits, and the widespread use and successful application of molecular biological techniques have made hybridization histochemistry much more generally acceptable. Still, the technique's importance only recently has been brought into sharp focus by applications from the latest studies in neuroscience, developmental biology, virology, and endocrinology. It now provides the precise cellular address of gene expression in functionally diverse and morphologically heterogeneous tissues, information that cannot be obtained by any other method.

The pursuit over the last 6 years of improvements in sensitivity, specificity, and resolution has resulted in publication of many varied procedures designed for different applications and of seemingly increasing complexity (see Coghlan et al., 1985, and other chapters in this volume). The choices of technique now available are based on various recent technological advances. The available probes include the familiar radiolabeled cDNA probes (Coghlan et al., 1985) as well as synthetic oligonucleotides (Coghlan et al., 1984; Coghlan et al., 1985), single-stranded cRNA (Cox et al., 1984) ("riboprobe"), nonisotopic cDNA probes (Brigati et al., 1983), and it is not too soon to speculate on the emergence of fluorescent probes.

While it is worthwhile to explore the possibility of a universal method, we are skeptical that it can be achieved. Each probe, with its own label, may have specific applications and procedures to optimize its individual use. Such aspects of the technique as the type of section, method of fixation, use and choice of proteinase, and type of probe and label may vary from laboratory to laboratory and person to person. The probe label determines the end-point, which may be autoradiographic or not, and thus the choice of film, emulsion, or method for detecting a nonisotopic label. These choices are influenced by whether precise cellular location is required vis-à-vis rapid screening, search of brain or whole mouse sections, or more precise quantitation of specific mRNA levels.

Our aim, based on extensive experience, is to provide some guidelines to simplify this choice. To this end we describe a convenient method to locate specific mRNA populations in brain and other tissues with either cDNA or synthetic oligodeoxyribonucleotide probes, which we favor.

METHODS

Probes

The full details of oligodeoxyribonucleotides synthesis have been published elsewhere (Penschow et al., 1986); they are based on the phosphoramidite technology of Caruthers (Caruthers et al., 1982; Adams et al., 1983).

End-Labeling of Oligodeoxyribonucleotides

We have observed that the use of a high pH, glycerol-containing buffer system (Procedure 5A in Maxam and Gilbert, 1980) gives consistently high

5'-end labeling with T$_4$ polynucleotide kinase and [γ-^{32}P]ATP. The levels of labeling achieved using this procedure are much higher, usually by a factor of 10, than the levels achieved using the procedure normally rec-ommended for the end-labeling of oligodeoxyribonucleotides (Procedure 5B in Maxam and Gilbert, 1980). The labeling is normally carried out on 100 ng of oligodeoxyribonucleotide (corresponding to 10 pmol of a 30 mer), using 20 pmol of [γ-^{32}P]ATP and 20 units of T$_4$ polynucleotide kinase, for 1 hour. The labeled probe is then purified on a Sephadex G-25 column (8 × 1 cm) precipitated in ethanol using 50 μg of tRNA to each fraction (8 drops each) containing the labeled product, dried under vacuum, and diluted to 400 ng/ml in hybridization buffer with 40% formamide. The specific activity ≃6 to 9 × 10^8 cpm/μg for a 30 mer (2–3 × 10^6 cpm/pmol). A small amount of the sample is checked for homogeneity by electrophoresis on a 10% polyacrylamide, 7 M urea gel. A discussion of 3' end-labeling using terminal deoxynucleotide transferase can be found elsewhere in this book (Chapter 7).

Labeling of cDNA Probes by Primed Synthesis

Although nick translation (Rigby et al., 1977) is used most widely, we find the random primer method simpler and more convenient (Taylor et al., 1976). Random primers 8 to 12 nucleotides in length are prepared by treating her-ring sperm DNA with DNase I and fractionating it on a DEAE-Sephadex G-50 column (Taylor et al., 1976). Primers (200 ng) and 100 ng of isolated insert DNA in 10 μl TE buffer (10 mM Tris, pH 7.5 with 1 mM EDTA) are boiled for 5 minutes and cooled on ice for 1 minute. This is incubated at 37° C in a final volume of 30 μl containing 20 mM of potassium chloride (KCl), 50 mM Tris HCl (pH 7.5), 6 mM magnesium chloride (MgCl$_2$), 1 mM of DTT, 500 μM each of dTTP, dGTP, dCTP, 50 μCi of [α-^{32}P]dATP (Amersham, >3000 Ci/mmol), and 5 units of DNA polymerase 1 (Klenow fragment). After 30 minutes, the reaction is terminated by the addition of EDTA to give a final concentration of 12.5 mM, and is extracted with phenol/chloroform (1:1). The aqueous phase is passed through a 10-×-60-mm col-umn of Sephadex G-50 (medium grade equilibrated with 0.1 M sodium chlo-ride [NaCl] in TE buffer). The labeled DNA probe, which elutes in the void volume, is ethanol precipitated and further treated as for synthetic oligo-deoxynucleotides. Specific activity is approximately 1 to 2 × 10^8 cpm/μg DNA. For probes of higher specific activity, [α-^{32}]dCTP can be used instead of unlabeled dCTP.

Procedure for Hybridization Histochemistry

The procedure is set out in a 15-step format outlined in Table 2-1, and described in the following detailed notes.

Table 2-1   *In situ* hybridization protocol

Cell cultures, cell smears	Fresh tissue specimens, plant or animal	Whole small animals
1. Fix in acetone and allow to dry at 4° C	Embed in OCT compound and freeze in hexane/dry ice	Freeze in hexane/dry ice, embed in 2% carboxymethylcellulose gel at 4° C and freeze-embed
2.	Cut sections 5–10 μm cryostat, thaw on to gelatinized slides, and leave on dry ice	Cut sections 40 μm in PMV cryomicrotome and collect on adhesive tape Photograph specimen between sections
3.	Fix in glutaraldehyde at 4° C	
4.	Rinse in hybridization buffer and "prehybridize" at 40° C	
5.	Rinse in ethanol, dry at room temperature	
6.	Dilute ^{32}P-labeled probe to 0.4 ng/μl in hybridization buffer and heat 3 minutes at 90° C (cDNA or oligonucleotides) or at 70° C (SP-6 RNA probes)	
7.	Apply probe to section and overlay with cover slip.	Lay section flat on glass plate, apply probe and overlay with 1.0-mm-thick glass plate
8.	Place in humidified chamber, cover with plastic film, and incubate at 30–42° C (DNA probes) or 50° C (SP-6-RNA probes) for 24 to 72 hours	
9.	Rinse slides individually in 4 × SSC, leave in 2 × SSC at room temperature, and wash in 1 × SSC at 40–50° C for 45 minutes	
10.	Rinse specimens twice in ethanol and allow to dry	
11.	Tape specimens to backing sheet in lightproof cassette	
12.	Count sections with Geiger counter; estimate exposure time	
13.	Apply fast x-ray film and expose at room temperature 6–24 hours, develop, fix, and wash	

Table 2-1 (*Continued*)

Cell cultures, cell smears	Fresh tissue specimens, plant or animal	Whole small animals
14. Estimate exposure time for liquid emulsion	Estimate exposure time for liquid emulsion and/or high-resolution x-ray film	Estimate exposure time for high-resolution x-ray film
15.	Dip slides in K5 emulsion; expose 1–14 days at room temperature, over silica gel; develop, fix, and wash	Apply film; expose for 1–14 days at room temperature; develop, fix, and wash.
16.	Stain and mount	Compare images with photographs of specimen or with adjacent sections

Step 1

Aluminum foil molds slightly larger than the specimen are used for embedding tissues. Accessory equipment for the PMV cryomicrotome includes molds for whole small animals.

Step 2

Slides are precoated in 1% gelatin hardened by rinsing in 0.25% formaldehyde, dried at 60 to 80° C, and used at room temperature. Freezing sections immediately on dry ice and leaving them for 30 minutes reduces ribonuclease activity and facilitates adherence of sections to slides. Adjacent sections retained for comparison with x-ray films are fixed for 10 seconds in 10% formaldehyde in 75% ethanol, and then are rinsed, stained, and mounted.

Because color photographs of the frozen cut surface of the animal show anatomic detail more clearly than adjacent sections, they are preferred as an aid to locating labeled areas on x-ray films.

Step 3

Glutaraldehyde fixative is 3% glutaraldehyde and 20% ethylene glycol in 0.1 M phosphate pH 7.2.

Step 4

Hybridization buffer is 600 mM sodium chloride, 50 mM sodium phosphate pH 7.0, 5.0 mM EDTA, 0.02% ficoll, 0.02% bovine serum albumin, 0.02% polyvinyl pyrrolidone, 0.1% herring sperm DNA, and 40% formamide deionized by adding 3 g/liter ion exchange resin (Bio-Rad, Richmond, CA, Catalog No. 142-67425) and filtered.

Step 5

At this stage, sections may be stored over ethanol at 4° C for 2 weeks or at −20° C for a few months with some risk of deterioration in morphological quality.

Step 6

Single-stranded probes with minimal self-complementary interactions are ensured by heat denaturation of labeled DNA probes at 90° C or RNA at 70° C immediately before application to sections. The hybridization signal increases with the period of incubation; 24 to 72 hours is a useful time frame, depending on the urgency and the desired signal intensity.

Step 7

The quantity of probe required depends on the size of the coverslip and size and thickness of sections. A rough guide is 20 μl of probe per 400 mm² (20 by 20 mm coverslip).

Step 8

The temperature of hybridization depends mainly on the probe length and degree of sequence homology between probe and target mRNA. Hybridizations of low-sequence homology or with short probes (less than 24 mer) are carried out at 30° C or lower whereas 50° C is more appropriate for RNA–RNA interactions, because of the stronger bonding, or for highly homologous DNA–RNA hybridizations. As adherence of sections to slides is reduced at higher temperatures, 40 to 42° C is preferred for most hybridizations with specific probes longer than 24 nucleotides.

Step 9

SCC in Table 2-1 stands for standard saline-citrate. Stock solution of 20 × SSC is 3 *M* sodium chloride and 0.3 *M* sodium citrate in distilled water.

The salt concentration and temperature of washing solutions are adjusted for removal of background "noise" while specific hybrids are retained. The effect of changes in temperature and salt concentration of washing solutions on the retention or removal of mRNA–DNA hybrids and background in tissue sections is, based on our experience, negligible compared with such effects on hybrids of DNA probes with extracted mRNA on nitrocellulose. 1 × SSC at 40° C is a useful standard; however, the temperature may be increased and salt concentration reduced for homologous and RNA–RNA interactions, and vice versa for short or less homologous hybrids.

Step 12

Counting sections with a Geiger counter provides an indication of whether the film can be developed within 24 hours.

Step 13

The fast x-ray film is developed for 2 minutes and fixed. Images obtained after 24-hour exposures should indicate whether or not the experiment was a success, and the signal intensity provides a guide to subsequent autoradiographic procedures and exposure time.

Step 14

The following comments refer only to ^{32}P-labeled probes, the procedures being dictated by the short half-life of this isotope (14 days). A weak signal on XAR-5 would indicate that liquid emulsion autoradiography (10–15 times slower than XAR-5) is futile so these specimens are usually exposed to high-resolution x-ray film (5 times slower than XAR-5) for a few weeks. "Hot" specimens may also be exposed in this way for a day or 2 and subsequently autoradiographed with liquid emulsion. Specimens with a moderate signal intensity are dipped directly in liquid emulsion. Exposure times for all the procedures are derived from XAR-5 images. There is little to be gained by exposures longer than the half-life of the isotope.

Step 15

The following emulsions are diluted 1:2 with distilled water and warmed for 3 h at 40° C.

> Liquid emulsion G5—exposure approximately 10 times XAR-5, crystal diameter 0.27 μm.
> Liquid emulsion K5–exposure approximately 15 times XAR-5, crystal diameter 0.20 μm.

The slides are dipped at 40° C; they are then exposed at room temperature over silica gel and, at 15° C, are developed for 2 minutes, rinsed in distilled water, and fixed.

The specimens are washed, stained, and mounted by conventional methods. Some special staining reactions may not succeed, probably because of the effects on tissue of formamide in hybridization buffer.

Examples of Technique

In using hybridization histochemistry, there is no substitute for viewing the labeled section through the microscope, using either light or darkfield. Color slides, taken with care, enable a larger audience to enjoy and appreciate the precision and delicacy of the technique. It must be pointed out, and is self-evident to those familiar with the procedure, that except where contact x-ray film is being used, conversion to black-and-white photos does not present the technique at anywhere near its best. In the symposium on which this volume is based, a wide range of oligonucleotide probes were used, partly to illustrate their versatility, but also to show that probe dilution is a useful quantitative approach and that suitably selected sequences could be made to separate closely related genes of the same family.

In our view, the best results are obtained by shaping the technique to the job in hand, and some examples have been selected to illustrate this point.

For hybridization of cells cultured on glass coverslips, the method must include a step that will alter membrane permeability sufficiently to allow the probe access to intracellular mRNA without causing mRNA to leach out of the cells. Minimal obvious damage to the cell membranes, and thus to cell morphology, is necessary for hybridization with short probes such as the ^{32}P-labeled 30-mer Ross River Virus probe (Dalgarno et al., 1983) shown in Plate 1a on cultured Vero cells. Cell morphology is clearly well preserved for the uninfected cell, whereas the probe has gained access to mRNA in the infected cell.

The location of gene expression in areas within a heterogeneous tissue or in single cells can be demonstrated quite adequately in autoradiographs of 5-μm frozen sections after hybridization with ^{32}P-labeled probes. Plate 1b further demonstrates the use of such end-labeled 30-mer oligodeoxyribonucleotides, where myocytes containing mRNA for atrial natriuretic peptide (Kangawa et al., 1984) are shown adjacent to an arteriole in the ventricle of the rat heart. Darkfield used for color and incident polarized illumination for black-and-white photomicrography, which show silver grains as white dots against the darker tissue background, are invaluable techniques for demonstrating labeled areas in autoradiographs, especially at low magnifications.

The collection and morphological preservation of surgical or field specimens for hybridization histochemistry may necessitate a portable tissue-freezing apparatus, comprising an appropriate solvent cooled by liquid nitrogen or dry ice, which can be set up in a nonlaboratory environment. The 5 μm frozen section of human placenta in Plate 1c was from tissue thus frozen by our standard method. The cDNA probe for human α-chorionic gonadotropin (Fiddes and Goodman, 1979) was labeled by primed synthesis with [α-^{32}P] dATP and has hybridized specifically to syncytial trophoblasts.

Intracellular granules, which are invariably lost from conventional frozen or paraffin sections, may be an important structural element in some cells and tissues, especially endocrine. A suitable example is the male mouse submandibular gland, shown in Plate 1d, where the mass of cytoplasmic granules of the granular convoluted tubule cells can be seen surrounding the lumen of the duct and ringed by the basally located nuclei. To demonstrate the granules, the tissue was frozen and then freeze dried, vapor fixed in paraformaldehyde, and vacuum embedded in paraffin; 3-μm sections were cut on a conventional rotary microtome. This method preserves mRNA as demonstrated here by the hybridization of a ^{32}P-labeled RNA probe generated from the 498-base pair (bp) cDNA for mouse glandular kallikrein (Richards et al., 1982) in the SP6 Riboprobe vector system (Melton et al., 1984).

X-ray film autoradiographs of hybridized tissues are useful, especially for screening for labeled areas in large sections. These x-ray pictures may also be used to compare relative levels of mRNA in tissue structures within a section. Although comparison by this method of mRNA levels in sections between animals is subject to some inaccuracies, it can provide useful semi-quantitative data in situations in which the accuracy of conventional quantitative techniques is impaired by problems with dissection or sampling. Comparison of x-ray films from a series of sections hybridized with labeled probe diluted with varying quantities of unlabeled homologous probe can provide more accurate information on relative mRNA levels in different areas within a section or in different tissues than a direct evaluation of films from single labeled sections. An example of this approach is shown in Figure 2-1, where sections from three experimental rat hearts have been hybridized with a ^{32}P-labeled 30-mer probe for atrial natriuretic peptide (ANP) mRNA (Kangawa et al., 1984). On each set of three sections (one from each animal) 10 ng of labeled probe was used, either undiluted or with 10 or 50 ng of unlabeled homologous probe added. The mRNA cellular levels for ANP in the ventricles of each heart are clearly lower than in the atria, as labeling in the ventricle is extinguished by $1\times$ probe dilution whereas differences in mRNA levels between the three atria become evident only at $5\times$ dilution. This shows the dehydrated rat atrium to have less ANP mRNA per cell than the normal or sodium-loaded rats, which were roughly equivalent, although a difference between these groups may be detectable with further probe dilutions.

Films can be processed by computer-assisted densitometry (Gouchee et al., 1980) to obtain more accurate measurements; however, sources of gross error, such as differences in section thickness or overexposed films, must be considered when data are being analyzed. Quantitative hybridization histochemistry provides information on the cellular labeling intensity only in the particular plane of section, and this information cannot necessarily be extrapolated to total mRNA levels in the whole piece of tissue or organ.

DISCUSSION

The hybridization histochemistry technique we have described has been applied successfully to botanical specimens (Anderson et al., 1986), whole mice (Coghlan et al., 1984; Rall et al., 1985), and cell smears and cultures (Plate 1a) as well as to a variety of tissues from humans (Zajac et al., 1986; and Plate 1c) and several different animal species (Coghlan et al., 1985).

Minor variations necessary for hybridization of cell cultures and whole mice have been included. This reliable method has produced reproducible positive results in our laboratory with at least 35 oligodeoxyribonucleotide probes ranging from 21 to 69 nucleotides and 25 cDNA probes ranging from 150 to 1700 bp (Table 2-2).

Table 2-2 Oligodeoxyribonucleotide and cDNA probes applied

Probe	Length (nucleotides)	Probe	Length (nucleotides)
Oligodeoxyribonucleotides			
r. Relaxin	30	m. β-Endorphin	
		Ross river virus	30
p. Relaxin	30	m. Atrial natriuretic peptide	30
h. Relaxin gene 2	24	m. Kallikrein	30
h. Prolactin	30	h. Calcitonin	40
h/m. α-Globin	30	h. Calcitonin gene related	
		peptide	35
o. β-Globin	30	r. Insulinlike growth factor II	30
o. γ-Globin	30	r. Insulin	30
b. Arginine vasopressin	30	m. Renin	30
b. Oxytocin	30	r. Growth hormone	30
r. Arginine vasopressin	30	m. Proliferin	30
r. Angiotensinogen	36	h. Parathyroid hormone	21
m. β Nerve growth factor	30	h. Kallikrein	69
m. α Nerve growth factor	30	o. Corticotropin-releasing	
		hormone	25
m. γ Nerve growth factor	30	p. α-Inhibin	60
		Stylar glycoprotein	
m. Renal kallikrein	30	(botanical)	30
m. Epidermal growth factor	30		
m. EGF binding protein			
Type B	30		
h. T-cell receptor	30		
cDNA			
b. Arginine vasopressin		h. Prolactin	
h. Factor 9		h. α-Chorionic gonadotropin	
h. Albumin		r. Growth hormone	
m. Insulinlike growth factor II		m. Kallikrein	
m. β-Nerve growth factor		r. Insulin	
b. Parathyroid hormone		m. Epidermal growth factor	
r. Calcitonin		m. Somatostatin	
h. Relaxin		h. Somatostatin	
p. Relaxin		m. Renin	
r. Relaxin		m. T-cell receptor	
m. β-Endorphin		m. Actin	
h. Proopiomelanocortin		o. Renin	
		Stylar glycoprotein	
		(botanical)	

Species: r = rat; p = porcine; h = human; o = ovine; b = bovine; m = mouse.

The critical first step, freezing the tissue, can determine the ultimate quality of the result. Some workers (Gee et al., 1983) prefer to first perfuse-fix and then freeze, but in many situations, such as the collection of human specimens, this is not feasible. A comparison of liquid tissue freezing methods (Elder et al., 1982) has shown that propane at around 83 K is the best general freezing method for optimal preservation of subcellular components for ultrastructural studies. The recommended maximum size of tissue samples is 3 mm³, which greatly limits the range of tissues frozen by this method that could be screened adequately by hybridization histochemistry. Tissue morphology in 5-μm frozen sections from specimens frozen in hexane cooled by solid carbon dioxide (CO_2) (dry ice) at −78° C compares favorably with similar sections of tissue frozen by the propane method. The size of specimens that can be frozen in hexane is limited only by the size of the freezing bath, although the quality of morphology is inversely related to specimen size.

A variety of fixatives have been reported as suitable for tissue hybridization, and some of these have been analyzed in detail on cell cultures as part of a method employing proteases after fixation and ^{32}P-labeled cDNA probes of 300 to 500 bp (Lawrence and Singer, 1985). The fixative recommended was 4% paraformaldehyde. A comparison we have made of morphology and signal intensity, using Carnoy's, Bouin's, and glutaraldehyde on 5-μm frozen sections hybridized with a 30-mer ^{32}P-labeled oligodeoxyribonucleotide, without protease treatment of sections, showed the signal intensity and background with Bouin's and glutaraldehyde to be equivalent. Glutaraldehyde provided superior morphological preservation while fixation with Carnoy's gave a low signal, the worst morphology of the three fixatives, and inconsistent staining reactions after fixation. Immersion fixation may be used for tissue hybridization (McAllister et al., 1983; Angerer and Angerer, 1981; Hafen et al., 1983) but is appropriate only for small specimens. It can be assumed that rapidly penetrating fixatives are essential to reduce possible degradation of mRNA by ribonucleases. Prolonged immersion in some fixatives has been reported to reduce mRNA levels (Brigati et al., 1983), which we also have observed. Because the cross-linking of proteins by aldehyde fixation may be reversible, storage of sections under ethanol prior to hybridization is recommended for optimal preservation of mRNA.

The actual hybridization procedures fall more into the realm of molecular biology than histology. The composition of hybridization buffer and the effects of temperature, salt, and formamide concentration on mRNA–DNA interactions have been well documented (Jones, 1973). Some authors have done rigorous checks on conditions (e.g., see Chapters 3, 4, and 11). Optimal conditions for hybridization histochemistry are not necessarily the same as for hybridization of mRNA immobilized on nitrocellulose. The concentration and specific activity of the probe needs to be somewhat greater than is generally employed for Northern gels or dot blots, and hybrids remain stable over a wider range of temperatures and salt concentrations for washing. The use of high temperatures (above 60° C) to increase the rate or specificity of hybridization may have a detrimental effect on sections or cause them to dislodge from slides. A comparison of some of the successful methods employed for hybridization histochemistry (Coghlan et al., 1985) indicates the wide variety of solutions and conditions in use. Additives, such as dextran sulfate (Wahl et al., 1979) to increase the signal, potassium iodide to reduce background for I^{125}-labeled probes (McAllister et al., 1983), and acetic anhydride (Hayashi et al., 1978) or herring sperm DNA for general background (Hudson et al., 1981), may not have beneficial effects for all methods and should be evaluated before inclusion.

Accurate estimation of exposure times for x-ray films depends mainly on experience but is also related to the type of film and isotope used, the concentration of the developer, and the temperature and duration of development. The rate of exposure of ^{32}P is increased at $-80°$ C with intensifying screens but with a reduction in the quality of resolution and tissue morphology. Exposure of a batch of liquid emulsion autoradiographs is less critical than x-ray film, because slides can be developed individually over a period to obtain a range of labeling intensity. The duration and temperature of development have less of an effect on signal intensity than grain size.

Labeling of cDNA probes, whatever the length, is surprisingly inefficient using either random primers or nick translations. Although it is satisfactory for many molecular biological applications, the specific activity level usually achieved with either of these procedures is just adequate for many applications we have tried but does not approach within more than a few percentage points or even less of the theoretical limit. Thus, probes labeled at 0.05% of theoretical or less, which is not uncommon, do not allow the hybridization histochemistry techniques to be operated at anywhere near maximum sensitivity. Because of the inefficiency of labeling, presumably because of competing chains, inefficiency of the enzymes, and the adequacy of the random primers, among other factors, end-labeling of synthetic 30 to 40-mer oligonucleotides gives specific activities that are, on the average, as high or higher.

Synthetic oligodeoxyribonucleotides have the advantages listed in Table 2-3. Because synthetic oligodeoxyribonucleotides can be labeled so efficiently, the method becomes liberated from a highly continuous dependence on the more complex techniques and know-how of molecular biology. This is not an insignificant matter for histology or pathology laboratories, which would like to use the technique. We have made general comments about the use of ^{32}P elsewhere (Coghlan et al., 1985), but other isotopes—^{35}S, ^{125}I, and ^{3}H—do have their place; ^{3}H, especially, improves resolution. Synthetic probes can be labeled with isotopic labels other than ^{32}P using different enzymes.

As mentioned before, the recently introduced SP6 Riboprobe system allows the production of considerable amounts of high-specific-activity complementary RNA from cDNA. This is a multistep process and, even in kits, requires expertise in molecular biology. There have been reports that labeled probes, although high in specific activity are of random length, often short. This may relate to the use of inadequate amounts of labeled nucleotides for

Table 2-3   Synthetic oligodeoxyribonucleotides—Advantages

Relative ease of manufacture in a few days
Can be efficiently labeled in an ordinary laboratory
Specific activities are more consistent (all probes from one batch of DNA)
A virtually unlimited number of sections can be hybridized (especially important when screening large sections)
Dilution experiments can be undertaken with unlabeled probes to provide data on relative mRNA levels (see Figure 2-1)
Synthetic probes give lower backgrounds
Shorter probes may be more accessible to cellular mRNA
Efficiency of hybridization is increased with single-stranded probes
Discriminating sequences for similar genes may be made
Known sequences may be used as models for different species using prepared codons
Attempts can be made to synthesize effective probes from the amino acid and sequence when the DNA sequence is still unknown

good enzyme efficiency. Where very high-specific-activity probes are required, the Riboprobe system clearly offers an alternative to labeled synthetic oligonucleotides. This has been a favored procedure for producing ^{35}S labeled probes, and difficulties with background are mentioned in Chapter 3, 6, and 11.

Synthetic oligonucleotides have disadvantages compared with other labeled probes, especially cDNA. These disadvantages are listed in Table 2-4.

One can take various approaches to quantitation. Evaluation of mRNA levels in relation to total tissue mRNA on northern gels can be made under similar or the same experimental conditions. Synthetic oligonucleotides allow a dilution to extinction type of approach to be used, as detailed earlier and in Figure 2-1. If x-ray film autoradiography is used, then various types of computerized densitometry can be employed. Color-coded image analyses can be a worthwhile adjunct (Mendelsohn et al., 1984). If ^{3}H-labeled probes are used, quantitation could be on the basis of grain counts over relevant cells. The physiological interpretation of the quantitation of mRNA levels from a single or a few 5-μm sections would, at best, have to be cautious. The heavy emphasis by referees and others on quantitating hybridization histochemistry micrographs in a way that simply does not occur with immunohistochemistry has intrigued us. Quantitation derived from *in situ* hybridization has its place, but sometimes the location of gene expression is all important.

Table 2-4   Synthetic oligodeoxyribonucleotides—Disadvantages

Some variation in hybridization conditions are required, especially with regard to probe lengths
If ^{32}P is used, then the section must be thin, sometimes giving a "worked out" appearance
Fewer labeling procedures are easily applicable
Not all probes complementary to mRNA work satisfactorily
Cross-species hybridization is more likely to be successful with longer probes
Errors in published sequence can be an occupational hazard

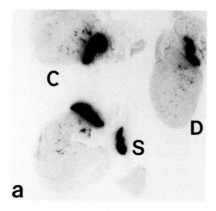
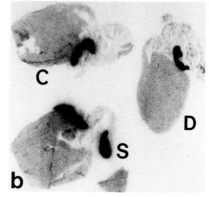

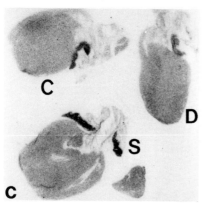

Figure 2-1 Autoradiographs on x-ray film of 5-μm frozen sections from hearts of three female Sprague-Dawley rats. D was hydrated (48 hours, without water), S was sodium-loaded (200 mM sodium chloride to drink for 7 days), and C was the control. Sections were hybridized with a ^{32}P-labeled oligodeoxyribonucleotide probe complementary to a 30-nucleotide sequence of the cardiodilatin region of mRNA for rat atrial natriuretic peptide: (a) 10-ng ^{32}P-labeled, 30-mer probe applied to sections; (b) 10-ng ^{32}P-labeled and 10-ng unlabeled homologous DNA probe applied to sections; (c) 10-ng ^{32}P-labeled and 50-ng unlabeled homologous probe applied to sections.

SUMMARY

This chapter has outlined a workable recipe for hybridization histochemistry, based on our recent experience. *In situ* tissue hybridization is an attractive and universally applicable research tool. Our recipe can be easily modified by any good "chef." We have not explored the conditions as rigorously as some others and have often resorted to the time-honored histological approach of trial and error. By the same token, we have tried to strip away as much empirical redundancy as we could. Although resolution down to a single cell in heterogeneous tissue was beyond the original expectation of the capability of ^{32}P-labeled probes, this can be achieved with care.

The main focus of the symposium on which this volume is based was applications of *in situ* hybridization to the nervous system; the technique offers much in this regard, given the many ways a peptide can enter a cell body. The technique has been very reproducible and satisfactory in its applications to the central nervous system, using relevant probes from Table 2-2.

Our introduction of the use of larger sections of brain and whole mouse sections in hybridization histochemistry has tremendous potential in locating inter alia hormonal, enzymatic and growth factor gene expression.

Improvements in probe labeling and in molecular biological procedures in the immediate future will no doubt make *in situ* tissue hybridization—hybridization histochemistry—a key technique in unraveling the complex processes of the central nervous system.

Sources of materials: Equipment and reagents

$[\gamma\text{-}^{32}P]$ ATP (5000 Ci/mmol) or higher (Amersham)
Herring sperm DNA (Sigma)
DNase (Sigma)
$[\alpha\text{-}^{32}P]dATP$ and $[\alpha\text{-}^{32}P]$ ($\approx$3000 Ci/mmol) (Amersham)
Deoxynucleotide triphosphates (dGTP, dCTP, dTTP) (Boehringer Mannheim)
Klenow (DNA polymerase) (B.R.E.S.A., Adelaide, South Australia)
"OCT" compound (Lab-Tek, Naperville, IL)
Cryostat-"Frigocut" (Reichert-Jung, Vienna, Austria)
PMV Cryomicrotome (L.K.B. Produkter, Stockholm)
Adhesive tape for sectioning—Scotch tape No. 688 (3M Co.)
Fast x-ray film—XAR-5 (Eastman Kodak Co., Rochester, NY)
Liquid emulsion: G5, K5, L4 (Ilford, Essex, England)
High-resolution x-ray film—Cronex MRF-32 (Dupont, Wilmington, DE)
For XAR-5 and MRF-32—Liquid x-ray developer type 2 and fixer (Kodak)
For liquid emulsions—D 19 developer (Kodak);—"Hypam" fixer diluted as directed (Ilford)
Riboprobe System (Promega Biotec, Madison, WI)

ACKNOWLEDGMENTS

Synthetic oligodeoxyribonucleotides were made in our laboratory or supplied by Syngene, in Melbourne, Australia.

Studies of computerized densitometry mentioned in this text are in progress in collaboration with Dr. F. A. O. Mendelsohn, University of Melbourne, Department of Medicine, The Austin Hospital.

This work was supported by grants-in-aid from the National Health and Medical Research Council of Australia; the Myer Family Trusts; the Ian Potter Foundation; and the Howard Florey Biomedical Foundation (United States); and by Grant No. HD 11908 from the National Institutes of Health.

REFERENCES

Adams, S.P., Kavka, K.S., Wykes, E.J., Holcher, S.B., and Galluppi, G.R. Hindered dialkylamino nucleotide phosphate reagents in the synthesis of two DNA 51-mers. *J. Am. Chem. Soc.* 105:661, 1983.

Anderson, M.A., Cornish, E.C., Mau, S.-L., Williams, E.G., Hoggart, R., At-
  kinson, A., Bonig, I., Grego, B., Simpson, R., Roche, P., Haley, J., Penschow, J., Niall, H., Tregear, G., Coghlan, J., Crawford, R., and Clarke,
  A.E. *Nature* 321:38, 1986.
Angerer, L.M. and Angerer, R.C. Detection of poly A$^+$ RNA in sea urchin eggs
  and embryos by quantitative *in situ* hybridization. *Nucleic Acids Res.* 9:2819,
  1981.
Brigati, D.J., Myerson, D., Leary, J.J., Spalholz, B., Travis, S.Z., Fong, C.K.Y.,
  Hsiung, G.D., and Ward, D.C. Detection of viral genomes in cultured cells
  and paraffin-embedded tissue sections using biotin-labeled hybridization probes.
  Virology 126:32, 1983.
Buongiorno-Nardelli, M. and Amaldi, F. Autoradiographic detection of molecular
  hybrids between rRNA and DNA in tissue sections. *Nature* 225:946 (Lon-
  don), 1970.
Caruthers, M.H., Beaucage, S.L., Efcavitch, J.W., Fisher, E.F., Goldman, R.A.,
  de Haseth, P.L., Mandecki, W., Matteucci, M.D., Rosendahl, M.S., and
  Stabinsky, Y. Chemical synthesis and biological studies on mutated gene-
  control regions. *Cold Spring Harbor Symposium. Quant. Biol.* 47:411, 1982.
Coghlan, J.P., Aldred, P., Haralambidis J., Niall, H.D., Penschow, J.D., and Tre-
  gear, G.W. Hybridization histochemistry. *Anal. Biochem.* 149:1, 1985.
Coghlan, J.P., Penschow, J.D., Tregear, G.W., and Niall, H.D. In *Receptors,
  Membranes and Transport Mechanisms in Medicine.* A.E. Doyle and F.
  Mendelsohn (Eds.). Excerpta Medica (Amsterdam), 1984, p. 1.
Cox, K.H., deLeon, D.V., Angerer, L.M., and Angerer, R.C. Detection of mRNAs
  in sea urchin embryos by *in situ* hybridization using asymmetric RNA probes.
  Dev. Biol. 101:485, 1984.
Dalgarno, L., Rice, C.M., and Strauss, J.H. Ross River Virus 26 S RNA: Complete
  nucleotide sequence and deduced sequence of the encoded structure. *Virology*
  129:179, 1983.
Elder, H.Y., Gray, C.C., Jardine, A.G., Chapman, J.N., and Biddlecombe, W.H.
  Optimum conditions for cryoquenching of small tissue blocks in liquid cool-
  ants. *J. Microscopy* 126(1):45, 1982.
Fiddes, J.C. and Goodman, H.M. Isolation, cloning and sequence analysis of the
  cDNA for the α-subunit of human chorionic gonadotropin. *Nature* 281:351,
  1979.
Gall, J. and Pardue, M. Formation and detection of RNA-DNA hybrid molecules
  in cytological preparations. *Proc. Natl. Acad. Sci. USA* 63:378, 1969.
Gee, C.E., Chen, C.-L.C., Roberts, J.L., Thompson, R., and Watson, S.J. Identification of proopiomelanocortin neurones in rat hypothalamus by *in situ* cDNA-
  mRNA hybridization. *Nature* 306:374, 1983.
Gouchee, C., Rasband, W., and Sokoloff, L. Computerized densitometry and color
  coding of [^{14}C] deoxyglucose autoradiographs. *Ann. Neurol.* 7:359, 1980.
Hafen, E., Levine, M., Garber, R.L., and Gehring, W.J. An improved *in situ* hybridization method for the detection of cellular RNAs in *Drosophila* tissue
  sections and its application for localizing transcripts of the homeotic *Antennapedia* gene complex. *EMBO* 2:617, 1983.
Hayashi, S., Gillam, I.C., Delaney, A.D., and Tener, G.M. Acetylation of chromosome squashes of *Drosophila melanogaster* decreases the background in
  autoradiographs from hybridization with [^{125}I]-labeled RNA. *J. Histochem.
  Cytochem.* 26:677, 1978.

Hudson, P., Penschow, J.D., Shine, J., Ryan, G., Niall, H.D., and Coghlan, J.P. Hybridization histochemistry: Use of recombinant DNA as a "homing probe" for tissue localization of specific mRNA populations. *Endocrinology* 108:353, 1981.

Jacobs, J., Simpson, E., Penschow, J.D., Hudson, P., Coghlan, J.P., and Niall, H.D. Characterization and localization of calcitonin messenger ribonucleic acid in rat thyroid. *Endocrinology* 113:1616, 1983.

John, H.A., Birnstiel, M.L., and Jones, K.W. RNA-DNA hybrids at the cytological level. *Nature* (London) 223:582, 1969.

Jones, K.W. In *New Techniques in Biophysics and Cell Biology*, Volume 1. R.H. Pain and B.J. Smith (Eds.). London: John Wiley & Sons, 1973, pp. 29–66.

Kangawa, K., Tawaragi, Y., Oikawa, S., Mizuno, A., Sakarugawa, Y., Nakazoto, H., Fukuda, A., Minamino, N., and Matsuo, H. Identification of rat γ atrial natriuretic polypeptide and characterization of the cDNA encoding its precursor. *Nature* 312:152, 1984.

Lawrence, J.B. and Singer, R.H. Quantitative analysis of *in situ* hybridization methods for the detection of actin gene expression. *Nucleic Acids Res.* 13:1777, 1985.

Maxam, A.M. and Gilbert, W. Sequencing end-labeled DNA with base-specific chemical cleavages. *Methods Enzymol.* 65:499, 1980.

McAllister, L.B., Scheller, R.H., Kandel, E.R., and Axel, R. *In situ* hybridization to study the origin and fate of identified neurons. *Science* 222:800, 1983.

Melton, D., Krieg, P., and Green, M.R. Functional messenger RNAs are produced by SP6 *in vitro* transcription of cloned cDNAs. *Nucleic Acids Res.* 12:7057, 1984.

Mendelsohn, F.A.O., Quirion, R., Saavedra, J.M., Aguilera, G., and Catt, K.J. Autoradiographic localization of angiotensin II receptors in rat brain. *Proc. Natl. Acad. Sci. USA* 81:1575, 1984.

Penschow, J.D., Haralambidis, J., Aldred, P., Tregear, G.W., and Coghlan, J.P. *Methods Enzymol.* 124:534, 1986.

Rall, L.B., Scott, J., Bell, G.I., Crawford, R.J., Penschow, J.D., Niall, H.D., and Coghlan, J.P. Mouse prepro-epidermal growth factor synthesis by the kidney and other tissues. *Nature* (London) 313:228, 1985.

Richards, R.I., Catanzaro, D.F., Mason, A.J., Morris, B.J., Baxter, J.D., and Shine, J. Mouse glandular kallikrein genes. *J. Biol. Chem.* 257(6):2758, 1982.

Rigby, P.W.J., Dieckmann, M., Rhodes, C., and Berg, P. Labeling deoxyribonucleic acid to high specific activity *in vitro* by nick translation with DNA polymerase I. *J. Mol. Biol.* 113:237, 1977.

Taylor, J.M., Illmensee, R., and Summers, J. Efficient transcription of RNA into DNA by avian sarcoma virus polymerase. *Biophys. Acta* 442:324, 1976.

Wahl, G.M., Stern, M., and Stark, G.R. Efficient transfer of large DNA fragments from agarose gels to diazobenzyloxymethyl-paper and rapid hybridization by using dextran sulfate. *Proc. Natl. Acad. Sci. USA* 76:3683, 1979.

Zajac, J.D., Penschow, J.D., Mason, T., Tregear, G.W., and Coghlan, J.P. *J. Clin. Endocrinol. Metab.* 62:1037, 1986.

3

In Situ Hybridization with RNA Probes: An Annotated Recipe

Lynne M. Angerer, Mark H. Stoler,
and Robert C. Angerer

Localization of individual mRNAs by *in situ* hybridization has emerged as a valuable tool in a broad range of research areas. Only immunological techniques and *in situ* hybridization allow identification of individual cells expressing specific genes. While these two techniques often provide similar information, the time or site of synthesis of a messenger RNA may differ from the time or site of accumulation of its protein (e.g., McAllister et al., 1983). Several factors make the *in situ* hybridization approach particularly appealing:

1. The specificity of the probe (a nucleic acid usually derived from a recombinant DNA clone) is inherent in its base sequence, and most sequences provide probes specific for single mRNAs. Even in the case of mRNAs transcribed from members of repetitive gene families, gene-specific probes can usually be acquired from 5' or 3' untranslated sequences that are less conserved than those encoding protein. Furthermore, nucleic acid probes can be developed even for highly conserved proteins that elicit antibodies only with difficulty.
2. Recombinant clones are essentially immortal and require little maintenance. One reasonably large preparation of a recombinant DNA can provide probes for years of *in situ* experiments.
3. Under suitable conditions, the *in situ* hybridization signal can provide quantitative estimates of relative, and some indication of absolute target mRNA concentration.
4. Although details of methods developed by different workers vary somewhat, the *in situ* technique is usually applicable to tissues fixed under conditions that provide good histological detail. Furthermore, fixed tissues are stable for long periods, allowing future comparisons or retrospective studies.

5. *In situ* hybridization methods have become increasingly sensitive. In the best cases, mRNAs present at the level of only a few molecules per cell can be detected. In fact, in the cases of mRNAs expressed in a small subfraction of a mixed population of cells, *in situ* hybridization may be much more sensitive than other nucleic acid hybridization techniques.

Several laboratories have contributed extensive analyses of the effects of different variables on *in situ* hybridization efficiency and optimized these variables for specific tissues. A number of refinements have contributed to the current sensitivity, characterization, and ease of application of this technique; however, we think three basic observations have been critical for increasing its sensitivity to a level permitting determination of the distributions of a large number of different mRNAs.

First, for mRNAs of typical length (about 1.5–2.0 kb), the ultimate signal depends on the fraction of target RNA that can be retained in tissues throughout the procedure. Although reports vary (see below), our opinion is that crosslinking fixatives such as formaldehyde and glutaraldehyde provide the highest potential signals in most cases.

Second, mRNA in tissue fixed with glutaraldehyde (and to a lesser degree, formaldehyde) is not readily accessible to probe molecules. Probe penetration is facilitated by use of proteases and probes of short fragment length (Brahic and Haase, 1978; Angerer and Angerer, 1981; Brigati et al., 1984).

Third, probes containing only the antisense strand usually provide considerably higher signals, because there is no sense strand to compete in solution for hybridization with target mRNAs *in situ* (implied by the data of Brahic and Haase, 1978, and Gall et al., 1981; demonstrated directly by Cox et al., 1984).

Our laboratory has investigated the use of antisense RNA probes because they offer a unique combination of advantages for *in situ* hybridization:

1. Lack of competing probe self-reaction, and therefore much higher signals.
2. Sequence purity. Single-stranded RNAs are transcribed *in vitro* from templates linearized with restriction endonucleases at sites just downstream from the inserted probe sequence. Because such transcripts include little contaminating vector sequence, they saturate target mRNAs *in situ* at lower overall concentrations than if vector sequence were also included. This affords a higher signal-to-noise ratio because nonspecific background binding of probe is proportional to the total concentration of labeled probe applied to the section.
3. Ease of preparation. Large quantities of RNA transcripts can be prepared by a relatively simple enzymatic synthesis followed by minimal purification. Fragment length of the probe is easily and reproducibly controlled by limited alkaline hydrolysis.
4. The probes can be labeled to high specific activity, equal to that of

the starting nucleotide precursors. In most cases, a relatively large fraction of the labeled precursors (40–80%) can be converted to RNA product.

5. Nonspecific backgrounds, in the form of probe sticking to tissue, can be easily and dramatically reduced by posthybridization digestion with RNase, under conditions in which probe–target hybrids are resistant.

6. The higher stability of RNA–RNA duplexes, compared with that of DNA–RNA, allows use of higher posthybridization wash temperatures to achieve a given fidelity of base pairing (stringency). Higher wash temperatures reduce nonspecific binding of probe to tissues.

7. Cellular transcription from most genes is asymmetric. DNA sequences inserted in transcription vectors in opposite orientations relative to the promoter encode either antisense or sense transcripts. Antisense transcripts detect mRNAs; sense transcripts provide controls for nonspecific background. Alternatively, sense strands provide specific probes for detecting DNAs, if these have first been denatured to permit hybridization to the probe.

The development of our protocol has been presented in two publications (Angerer and Angerer, 1981; Cox et al., 1984), and we have recently reviewed most of the variables in this procedure (Angerer et al., 1985a). Here we will provide detailed recipes for individual steps and emphasize several important aspects of the technique.

TISSUE PREPARATION

Preliminary Comments and Cautions

Much of the diversity of report, opinion, and practice of *in situ* hybridization revolves around procedures for fixing tissue, and for prehybridization treatments designed to increase access of probe to target mRNAs. Many apparent differences can be rationalized if the function of individual steps in the procedure and their interrelationships are understood. It is most important to note two facts. First, tissues are different. Because their biochemical composition is highly variable, they may respond differently to fixatives and treatments designed to increase probe penetration and may also be subject to variable levels of adventitious binding of probe. There is, therefore, no reason to expect that one procedure will be the best for all tissues. Second, it is very important to recognize that individual steps in any protocol are interdependent. Some procedures that work very well in the context of one protocol are *disastrous* in different combinations. Some of these interactions are discussed in the following paragraphs.

Table 3-1 lists three common fixatives that have been used for *in situ*

Table 3-1 Comparison of three common fixatives

	EtOH/HAc	Formaldehyde	Glutaraldehyde
Morphology	Fair	Good	Good
RNA retention	Low	Good	Very good
Protease helps?	No	Usually	Definitely

hybridization—ethanol/acetic acid (EtOH/HAc), formaldehyde (or para-formaldehyde), and glutaraldehyde. These are arranged in order of increasing "tightness" of the tissue after fixation, which appears to parallel the retention of cellular RNA in sections. There is probably a general inverse relationship between RNA retention and accessibility of target RNAs to hybridization probes. Aldehyde fixatives provide better retention of cellular RNA (except, perhaps in the case of very long viral RNAs, which are more easily retained), but deproteinization is usually required to take advantage of this potential for higher signals. Choice of fixative and prehybridization treatment provides one major example of interacting variables: We found that although proteinase K treatment increased signals about 10-fold with glutaraldehyde-fixed material (Angerer and Angerer, 1981), signals were lost when this pretreatment was applied to EtOH/HAc-fixed material, presumably because in the latter case proteinase K treatment released target RNAs from sections.

We routinely use 1% glutaraldehyde to fix sea urchin embryos and tissue, as described later, and our measurements indicate that most of the target mRNA is retained and accessible to hybridization probes after appropriate protease treatments (Cox et al., 1984). This fixative has also been successful with pellets of cultured cells (mouse myeloma) and whole unicellular eukaryotes (e.g., *Tetrahymena;* S.-M. Yu and M. Gorovsky, personal communication). Many other workers use formaldehyde as a fixative, which is also successful for many tissues in our hands (see, for example, Plate 2), although we have not directly compared signals obtainable with formaldehyde and with glutaraldehyde. Use of tissues prepared by standard formalin fixation procedures has the practical advantages of allowing retrospective studies of large archives of clinical specimens and facilitating prospective studies. However, investigations involving quantitative estimates of signals will require development of careful methods of standardization for differences in magnitude of signals for tissues fixed after different postmortem intervals, for different times, and so on.

Fixation

Glutaraldehyde

1. Suspend cells or small tissue fragments in at least 5 volumes of ice-cold 1% glutaraldehyde, 50 mM sodium phosphate buffer, pH 7.5,

and sufficient NaCl to give an osmolality of fixative that avoids excessive shrinking or swelling. The osmolality of 1% glutaraldehyde is approximately equivalent to 0.375% NaCl (Millonig and Marinozzi, 1968).

2. Remove fixative, add fresh cold fixative, and incubate on ice for a total of 1 hour.

3. Wash by resuspending tissue in at least 10 volumes of buffer at 0° C for 30 minutes. Repeat the procedure for a total wash time of 1 hour.

Formaldehyde

Tissue sections 1 to 3 mm thick are fixed in 10% neutral buffered formalin (4% formaldehyde, 0.4% NaH_2PO_4, 0.65% Na_2HPO_4, 1.5% methanol, all weight/volume, pH 7.0). Tissues are fixed as soon as possible after removal, to minimize degradation of nucleic acids. Typically, fixation is for 1 to 6 hours at room temperature. Thicker tissue sections require longer fixation times, while cell suspensions can be fixed in as little as 30 minutes. In the few cases examined so far, we have not detected any adverse affects of fixation for periods as long as several days.

Our experience with formalin-fixed specimens has so far been limited to tissues that were subsequently processed routinely and paraffin-embedded using an automated vacuum infiltration processor. Several different tissues (brain, cervix, skin, breast, thyroid), and cell types, including archival material, have been used for hybridization with a variety of probes. Our experience to date suggests that tissue prepared by routine histological processing is suitable at least for qualitative identification of cells expressing individual mRNAs.

Notes. As discussed for formaldehyde, appropriate times for fixation and washing in any protocol will vary with the specimen size.

Embedding in Paraffin and Sectioning

Dehydration

1. Wash one time with 0.5 M NaCl to remove the phosphate (which would precipitate in ethanols), while maintaining ionic strength. Resuspend the tissue in at least 10 tissue volumes of the same solution and add 99% EtOH while stirring to a final concentration of 50% EtOH at room temperature. After 15 minutes, let the tissue settle and resuspend it in fresh 50% EtOH, 1.25% NaCl, for 15 minutes. Decant.

2. Resuspend in 70% EtOH at room temperature. Decant and add fresh 70% EtOH after 15 minutes.

3. Dehydrate the tissue in the following series of solvents at room tem-

perature for 30 minutes each, using at least 10 tissue volumes: 85%, 95%, 99%, and 99% EtOH; xylene, xylene.

Paraffin Embedding

1. Infiltrate the tissue with paraffin as follows: Xylene: paraffin (1:1), 45 minutes at 58° C. Paraffin, three changes of 20 minutes each, at 58° C.
2. Small pieces of tissue are easily placed in molds. For embedding small embryos (i.e., less than 0.5 mm), transfer a drop of a concentrated suspension quickly to an embedding mold using a prewarmed wide-bore pasteur pipet. Allow the drop to solidify and then place it in an oven at 58° C just until the surface begins to glisten. Immediately add sufficient liquid paraffin to cover the drop. Allow the mold to sit at room temperature until a skin forms, then fill it with liquid paraffin and allow it to solidify at room temperature. An alternative method for small quantities of tissue is to embed in molds used for plastic embedding (BEEM capsules), or in microcentrifuge tubes. One way of conveniently handling cell suspensions is to first embed them in agarose by mixing concentrated fixed cells with an equal volume of 1.6% low-melting-temperature agarose.

Notes. It should be emphasized that the times given here are for small pieces of tissue or cell suspension and should probably be increased in proportion to the size of the tissue blocks.

We have found initial dehydration steps to be critical, and the procedure given was developed largely by trial and error. Initial dehydration steps that use too low EtOH and/or salt concentration result in irreversible expansion and destruction of the tissue. The high salt concentrations given here are appropriate for a marine invertebrate embryo, but are not necessarily the best for other tissues. When in doubt, use higher ethanol and/or salt concentrations for the initial dehydration steps; at worst, these cause shrinkage.

The temperature must be closely controlled during the paraffin infiltration; this is done most conveniently by using a heat block. We use paraplast with a melting temperature of 55 to 57° C. Overheating the paraffin (>60° C) damages its properties, and sections fragment during spreading. The properties of paraplast also change on prolonged storage when melted. Melt the paraffin the night before it is used, and before embedding degas liquefied paraffin in a vacuum flask (or melt it overnight in a vacuum oven), to prevent bubble formation.

It is possible to store tissue in paraffin in test tubes at 4° C, and then melt and embed it at a later time. This allows mixing of different samples of cells or small pieces of tissue in the same block, which affords the best comparison of relative signals. Tissue is stable in paraffin for at least several years without noticeable decrease in signals or deterioration of morphology. Fixed tissue may also be stored in 70% EtOH at 4° C for months, although it does not appear to be as stable as in paraffin.

Slide Preparation

1. Cleaning. Slides are stored indefinitely or at least overnight in Chromerge, and then rinsed at least 15 minutes each in running tap water and running distilled water. For these washes it is convenient to insert slides vertically in a test tube rack (40 slides per rack). Finally, rinse the slides individually by holding them under a forceful stream of distilled water. Handle them only by one end, with forceps. Air dry, in a dust-free place.

2. Polylysine coating. This is a modification of the procedure of McClay et al. (1981). Immerse the slides for 10 minutes at room temperature in slide mailers (to conserve reagent) filled with 50 µg of poly-L-lysine per milliliter, 10 mM Tris HCl, pH 8.0. (A 10× stock solution can be stored frozen, indefinitely.) At least four sets of slides can be coated from each volume of polylysine (just sufficient to cover the slides). Air dry in a dust-free place. The slides will have spots of polylysine on them. Coated slides keep for at least several weeks at room temperature.

Notes. We had considerable difficulty with loss of sections or parts of sections from slides during the hybridization procedure using slides coated by a variety of procedures, including gelatin (Gall and Pardue, 1971), egg white, or "Histostik" (Brigati et al., 1984). The polylysine method has proved far superior to the others. Although it might have been expected that polylysine would bind probe, and thereby increase backgrounds, we have not had such a problem, possibly because the slides are treated with proteinase K and acetic anhydride before probe is applied.

It has been reported that retention of sections increases with increasing molecular weight of the polylysine (Huang et. al., 1983).

Sectioning

Sectioning is much easier to learn by observation and a few hours in an histology laboratory is probably a good investment. Following is the basic procedure.

1. Mount the paraffin block by pressing its backside on a heated chuck or holder.

2. With a razor blade, trim the block into a truncated pyramid with a trapezoidal face, slightly wider at the bottom than at the top of the section, so that the leading edge of each section is slightly wider than the tailing edge of the previous one, which it displaces from the knife blade. Trim away as much paraffin as possible without losing the tissue. Smaller blocks are easier to section, and are thus better for serial sections.

3. Section at a nominal thickness of 5 μm. Sectioning problems can usually be attributed to dull blades or poor quality paraffin.

4. Ribbons of sections can be picked up by one end using a small spatula wet with a drop of water and are then floated on 45° C degased distilled water until they spread to the size of the original block face. This should take only a few seconds. (Water must be degased to avoid formation of bubbles between the sections and the slide, and consequent loss of parts of the sections.) Pick up the sections on coated slides; handling the slide only by the edges at one end, insert it in the water near the sections at a 45° angle. Approach the sections slowly until their near edge touches the slide, and slowly withdraw the slide from the bath. Keep the sections between about 1 and 5 cm from the left edge of the slide so that they are sure to be covered by emulsion during dipping, but are not too close to the bottom where the autoradiographic emulsion is thicker and drying artifacts may occur.

5. Incubate the slides on a slide warmer at 40° C for 2 hours. Shorter times may lead to more loss of sections. We routinely incubate large sections of formalin-fixed tissues at 60 to 62° C for 45 minutes to 1 hour, with the slides placed vertically.

6. Store the slides in a clean dry slide box at room temperature.

Notes. We have little experience with storing paraffin sections for periods longer than several weeks, and usually use them within a few days.

It is also possible to cut 1-μm sections with glass knives on an ultramicrotome. Because adjacent sections pass through essentially the same set of cells, they are particularly useful for comparing distributions of different mRNAs using two different probes hybridized to adjacent sections (for example, see Angerer et al., 1985b). Hybridization of adjacent sections with the same probe demonstrates the reproducibility of patterns, especially if signals are low, and helps distinguish between experimental and biological variability. Thinner sections also facilitate demonstration of nuclear labeling, since they reduce the frequency with which nuclei are covered by an overlying layer of cytoplasm that may either quench a true nuclear signal or provide a false nuclear signal that is actually a result of cytoplasmic RNA. Finally, thin sections allow more experiments from limited quantities of tissue.

Prehybridization Treatments

Deparaffinization and Hydration of Sections

1. Place the slides in glass slide carriers and immerse them in two changes of fresh xylene, each for 10 minutes. Either shake gently on a shaking platform, or stir gently using a magnetic stirbar under the slide rack.

2. Pass the slides sequentially for about 15 seconds each (dipping up and down about 10 times) through 99%, 99%, 95%, 85%, 70%, 50%, and 30% EtOH, water, water.

Proteinase K Digestion

Immerse the slides in 100 mM Tris HCl, pH 8.0, 50 mM EDTA, at 37° C, and 1 μg of proteinase K/ml for 30 minutes. The buffer is prewarmed and the enzyme added from a 10-mg/ml stock just before it is used.

Acetic Anhydride

The purpose of this step is to acetylate amino groups in tissues and thus reduce electrostatic binding of probe (Hayashi et al., 1978). In our particular sequence of steps, the acetic anhydride may also serve to block potential binding of probe to polylysine and to inhibit proteinase K activity.

1. Wash the slides briefly in distilled water and then in freshly prepared triethanolamine HCl, pH 8.0, at room temperature. Blot the slide carrier on paper towels to remove excess buffer.
2. Add sufficient undiluted acetic anhydride to an empty dry staining dish to give a final volume of solution that will cover the slides and give a final concentration of 0.25% (Volume/Volume). Place the rack of slides in the dish and immediately add 0.1 M triethanolamine, pH 8.0, to the final volume. Quickly dunk the slides up and down several times to mix in the acetic anhydride. Incubate at room temperature for 10 minutes.
3. Wash briefly in 2 × SSC.

Dehydration of Sections

Dehydrate the sections by passing the slides through the graded ethanol series in reverse order up to 99% EtOH. (The ethanols used for dewaxing the slides may be reused for this step.) Place the slides vertically in a test tube rack on a paper towel to dry. Cover them to keep them dust-free.

Modification for Detection of DNA

After proteinase K and acetic anhydride treatments, denature the target DNA by immersing the dehydrated slides in 95% formamide, 0.1 × SSC for 15 minutes at 65° C. Transfer the slides to ice-cold 0.1 × SSC for 5 minutes, dehydrate, and continue as for mRNA hybridization. We do not know what fraction of DNA sequence is made available for hybridization by this treatment.

Notes. The indicated proteinase K treatment has been observed to be optimal in several different systems. However, the extent of digestion required may vary with tissues and fixation conditions. For example, signals continue to increase for whole-mounted *Tetrahymena* fixed in 1% glutaraldehyde up to 30 μg of proteinase K per milliliter without significant damage to morphology (S.-M. Yu and M. Gorovsky, personal communication), although for sea urchin embryos considerable deterioration is observed above 5 μg/ml, without an increase in signals. A range of enzyme concentrations should be examined to find an optimum that maximizes accessibility of target RNAs as much as possible without causing loss of RNA from sections, or deterioration of morphology. Target DNA (e.g., viral genomes) is usually high molecular weight and is probably more easily retained in sections. In this case, higher proteinase concentrations may be advantageous.

Those who use formalin-fixed material often suggest postfixation after protease treatment to improve RNA retention (Haase et. al., 1982; Brigati et al., 1983). This has not been required with glutaraldehyde fixation, presumably because the RNA is more tightly cross-linked. Brigati et al., (1983) have favored pronase in combination with formaldehyde-fixed material, largely because pronase activity can be blocked with glycine. We prefer proteinase K, originally suggested by Brahic and Haase (1978), because of its reproducible activity, stability (up to 1 year in 10 mg/ml stock, frozen), and freedom from nucleases. Proteinase K activity is probably terminated immediately by subsequent acetic anhydride treatment.

PROBE PREPARATION

Procedures for *in vitro* transcription of RNAs are discussed in more detail by Melton et al., (1984), Krieg et al. (1985), and Angerer et al. (1985a).

Transcription Vectors

The original transcription vectors contained a promoter from the *Salmonella* phage, SP6, and synthesis of both sense and antisense strand transcripts required separate clones containing the inserted probe sequence in opposite orientations with respect to the SP6 promoter. More recent commercially available vectors carry two different apposed promoters (SP6, T7 or T3) separated by a multiple cloning site, into which the probe sequence is inserted. The specificity of each polymerase for its promoter allows synthesis of RNAs representing either strand of the insert from the same clone. All the enzymes are commercially available and have similar properties. The T3 and T7 polymerases are currently cheaper than SP6 polymerase, but isolation of large quantities of the latter enzyme is relatively simple and has been described by Butler and Chamberlin (1982) (see Cox et al., 1984, for two typographical corrections of the original published protocol).

Preparation of Template

1. Purify the DNA. For obtaining small quantities of plasmid DNA sufficiently pure to support good transcription *in vitro,* we have found the alkaline lysis procedure of Krieg and Melton (1985) to be the most reliable. Larger quantities of plasmid DNA can be prepared by standard methods using cesium chloride (CsCl) equilibrium centrifugation (Maniatis et al., 1982).
2. Truncate the template (250–500 μg/ml) by digestion of the circular supercoiled molecules with a restriction endonuclease that cuts just downstream of the insert sequence, but not elsewhere between the promoter and the desired truncation site.
3. Remove the buffer by dialyzing the DNA at room temperature for 30 minutes versus 5 mM Tris HCl, pH 8.0. This is conveniently done by placing the sample (25–100 μl) on a Millipore VMWP filter, which floats on the dialyzate with the shiny (hydrophobic) side up. Significant increases in volume may occur, which should be taken into account when estimating the final DNA concentrations.

In Vitro Transcription

1. Final concentrations are 40 mM Tris HCl, pH 7.5, 6 mM MgCl$_2$, 10 mM dithiothreitol (DTT), 100 to 200 μM each of XTP, 2 mM spermidine, about 100 μg/ml of truncated template DNA, 100 units of placental RNase inhibitor/ml, and 1200 to 1800 units of SP6 RNA polymerase per milliliter.
2. Assemble the mix as follows:
   a. Dry down the radioactively labeled precursor(s) in a microcentrifuge tube under vacuum.
   b. Dissolve the labeled nucleotides in a solution containing the Tris, MgCl$_2$, DTT (made fresh), cold XTPs, RNase inhibitor, spermidine, and water (as necessary).
   c. Add template to the side of the tube, and mix it in quickly to avoid local high concentrations, which might lead to precipitation of DNA by the spermidine.
   d. Add enzyme.
3. Incubate at 37 to 41° C for 10 minutes to 2 hours or more, depending on the template length; further additions of enzyme may be advantageous (see later). Incorporation is assayed by acid precipitation.

Notes. Truncated templates encode transcripts of the sequence between the promoter and the truncated end, avoiding transcription of vector sequence.

It is important to be sure that templates are completely linearized with restriction endonuclease, especially if the template is short. Even small amounts of uncut supercoils result in disproportionately high fractions of RNA mass in long transcripts containing vector sequence because initiation is rate limiting, and the mass of transcript per initiation is much higher for circular templates.

In enzyme excess the rate of incorporation of nucleotides is approximately proportional to the length of the template sequence. Short templates require longer times to achieve a given level of precursor incorporation. Only long templates (>several kilobases [kb]) support incorporation rates approaching those indicated by manufacturers' units, since the unit is defined on a supercoiled template. Consequently, short templates require longer synthesis times for high levels of incorporation.

Inclusion of RNase inhibitor helps prevent degradation of the product. This is especially important if the DNA template purification has used RNase A digestion to remove bacterial RNA. A small amount of degradation is not harmful, since the fragment length of the probe transcripts is intentionally reduced before hybridization. However, it is wise to check the length of RNA transcripts from new templates by electrophoresis on denaturing agarose gels (see later) to be sure that the template does not contain strong transcription termination sites.

It is important to use nucleotide triphosphate concentrations well above the K_m of the enzyme (which are 10–60 μM, depending on the nucleotide; Butler and Chamberlin, 1982), because at very low precursor concentrations or high concentrations of template sequences, or both, only promoter–proximal portions of the insert sequence may be transcribed before precursors are depleted. This reduces the potential hybridization signal proportionately. We normally use at least 100 μM for the radioactive precursors and 200 to 400 μM for unlabeled nucleotides.

For plasmids 4 to 5 kb in length, the DNA concentration is approximately 100 $\mu g/ml$; for longer plasmids, the template concentration is adjusted to maintain approximately the same promoter concentration (about 0.4 pmol of promoter per 10 μl of synthesis). We use about 10 units of polymerase per microgram of 4- to 5-kb DNA, which normally achieves slight enzyme excess.

The SP6 polymerase requires clean templates. If some preparations do not support high rates of synthesis, then further purification (phenol, chloroform, ethanol precipitation) may be helpful.

Probe Purification

1. Remove DNA template. Dilute the polymerase reaction tenfold by adding DNase I buffer (50 mM Tris HCl, pH 7.4, 10 mM MgCl$_2$) and incubate with 50 μg of RNase-free DNase I per milliliter for 30 minutes at 37° C.

2. Extract with phenol/chloroform, chloroform, and precipitate with ethanol. Alternatively, precipitation with 2 *M* of ammonium acetate (NH_4Ac) and 2.5 volumes of ethanol gives better removal of unincorporated XTPs (Maniatis et al., 1982). Our probe preparations are usually greater than 80% acid precipitable when used for hybridization.

Reduction of Probe Fragment Length

This step is included because shorter probes give higher hybridization signals (Brahic and Haase, 1978; Angerer and Angerer, 1981).

1. Dissolve the probe in 50 μl of water and add to it an equal volume of 0.2 *M* carbonate buffer, pH 10.2 (80 m*M* $NaHCO_3$/120 m*M* Na_2CO_3). The buffer is made fresh or stored frozen.
2. Incubate for the appropriate time at 60° C. Hydrolysis time is given by the following relationship (Cox et al., 1984):

$$t = \frac{L_0 - L_f}{kL_0L_f}$$

where

t = time in minutes
L_0 = initial length in kilobases
L_f = desired length in kilobases
k = approximately 0.11 strand scissions/kilobase/minute

3. Neutralize by adding 3 μl of 3*M* NaAc, pH 6.0, and 5 μl 10% (Volume/Volume) glacial acetic acid. Precipitate with 2 volumes of EtOH.
4. Resuspend the dried precipitate in a small volume of water and store it frozen. ^{35}S-labeled probes should be resuspended in 20 m*M* DTT.
5. Determine the size of hydrolyzed probes on 2 to 2.5% agarose gels in MOPS buffer (20 m*M* MOPS, 5 m*M* NaAc, 1 m*M* EDTA, pH 7.0), containing 6% formaldehyde, as described by Bruskin et al. (1982). Suitable radioactive size markers can be prepared by transcription from a mixture of templates prepared by truncating a vector containing an SP6 promoter with appropriate restriction endonucleases.

Notes. We use probes whose mass average size is about 150 nt. This is short enough to provide good penetration into the tissue, but not so short that there is a significant reduction in hybrid T_m, or much variation in T_m as a function of minor differences in fragment length (see later).

Typical hydrolysis times are in the range of 30 to 45 minutes. Because k is relatively small and fragment length is an exponential function of hydrolysis time, control of the fragment length is rather reproducible.

HYBRIDIZATION

Preparation of the Hybridization Mix

1. Final concentrations are 50% formamide, 0.3 M NaCl, 20 mM Tris HCl, pH 8.0, 1 mM EDTA, 1 × Denhardt's (0.02% each BSA, ficoll, polyvinyl pyrrolidone), 500 μg of yeast tRNA per milliliter, and 10% dextran sulfate. Add 500 μg of poly-A per milliliter for cDNA probes that contain poly-U tracts. For ^{35}S-labeled probes, include 100 μM DTT.
2. Prepare a mix of sufficient volume to use for all the slides, containing the formamide, NaCl, Tris, EDTA, Denhardt's, and dextran sulfate.
3. Combine the probe, tRNA, and poly-A in low salt (5 mM Tris HCl, pH 8.0) with sufficient water to achieve the correct final volume, and heat it, at 80° C for 3 minutes to ensure that all the nucleic acids are disaggregated. Quick cooling is not required.
4. Combine the probe and the mix from step 2 at room temperature.

Setting up the Slides

1. To pretreated slides, which are completely dry, add hybridization mix (3–4 μl/cm^2) at the center of the area containing sections. Use of dry slides ensures that the concentrations of probe and reagents in the hybridization solution are well controlled.
2. Cut siliconized, baked (150° C, 2 hour) coverslips to maximize coverage of the sections and minimize the probe required. Coverslips are cut by scoring with a diamond pencil on a hard, flat surface.
3. Using fine-tipped forceps, apply one edge of the coverslip to the drop of probe mix and slowly decrease the angle between the coverslip and the slide so that bubbles will be forced out toward the other end. Coverslips and slides must be very clean to minimize the bubbles. Do not press down on the coverslips, because this results in a thinner layer of probe solution and decreased signals. The viscosity of the probe solution (due largely to the dextran sulfate) and the uniform weight/area of the coverslip give a reproducible thickness of probe solution.
4. If not all sections are completely covered with probe, mark the position of the coverslip on the underside of the slide with a diamond pencil. Any large bubbles can also be marked. (The probe mix holds the coverslip in place, so that the slide can be turned upside down.)
5. Immediately, immerse the slides in mineral oil in a covered plastic box and incubate at the appropriate temperature. (Mineral oil is messy, but provides a foolproof seal. The oil can be reused indefinitely.)

Choosing Probe Concentration, Hybridization Time, and Temperature

Probe Concentration and Hybridization Time

The kinetics of hybridization *in situ* are discussed in Cox et al. (1984) and Angerer et al. (1985a). The fraction of maximal signal obtained depends on probe concentration and time, but not reciprocally, even through the probe is theoretically in large excess. Thus, much of the probe must not participate in the *in situ* reaction. Although diffusion is expected to be limited, the high levels of nonspecific binding of probe to tissue in the absence of posthybridization RNase treatment (our unpublished observation) suggest that some probe is also removed from the reaction by adventitious binding to tissue. The extent and time course of hybridization are very similar using slides coated with polylysine, or with gelatin-chrom-alum (Figure 3-1) as originally used by Cox et al. (1984). Although Cox et al. (1984) showed that hybridization for a given probe concentration is largely completed after about 6 hours, a recent examination of longer hybridization times (Figure 3-1) demonstrates a slow second phase of hybridization. The additional signal achieved however, is rarely worth extending hybridization beyond overnight (16 hours), which is our standard.

Using this fixed time, signals increase linearly as a function of probe concentration until saturation is achieved (Cox et al., 1984). In our system saturation requires 0.2 to 0.3 μg of probe per milliliter per kilobase of probe complexity (i.e., total RNA transcript sequence complexity, whether or not the sequence is complementary to cellar RNAs). Maximal signal-to-noise ratios are achieved at probe concentrations just sufficient to saturate target RNAs; higher concentrations only increase backgrounds. Probes of higher

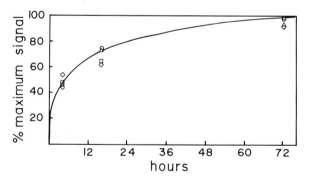

Figure 3-1 Kinetics of hybridization *in situ* with single-stranded RNA probes. A probe transcribed from an *S. purpuratus* early histone gene repeat sequence was hybridized to sections of 12-hour embryos, as described by Cox et al. (1984). The probe concentration was 0.4 μg/ml (□,△) or 2.0 μg/ml (○,◇). Absolute signals were higher for the higher probe concentration, and the data have been normalized to the same final value for comparison. Tissue sections were mounted on slides coated with polylysine (□,○) or chrom-alum (△,◇). Note that the rate of hybridization is independent of mounting substrate and that the apparent rate of hybridization is similar, although probe concentrations vary five-fold.

sequence complexity complementary to target mRNAs provide proportionately higher signals at saturation, as well as proportionately higher backgrounds, since higher probe concentrations are required to achieve saturation. Thus, longer probes require shorter exposures to achieve the same signal,
but do not result in higher signal-to-noise ratios.

Inclusion of dextran sulfate increases hybridization signals about fivefold
(see Figure 3-2). Although dextran sulfate is known to increase nucleic acid
reassociation rates through an excluded volume effect, it may also help prevent adsorption of probe from the reaction by competing for adventitious
binding to tissue.

Hybridization Stringency, Optimum Temperature, and the Specificity of Base Pairing

A number of factors affect the appropriate temperature for hybridization,
which in turn determines the rate and specificity of the reaction. The ap-
porpriate temperature is calculated by reference to the T_m of the duplexes
formed. An equation for calculating T_m values of RNA–RNA duplexes
(compiled in part by Bodkin and Knudson, 1985) is

$$T_m = 79.8 + 58.4 \, (F_{GC}) + 11.8 \, (F_{GC})^2$$

$$+ \, 18.5 \log(M) - \frac{820}{L} - 0.35(\%F) - (\%m)$$

where G is guanosine, C is cytosine, A is adenosine, and T is thymidine.

F_{GC} is the mole fraction GC content; these terms reflect the higher stability
of GC over AT base pairs. If the GC content of a probe is an unknown, 40
to 50% is usually close.

M is the monovalent cation concentration. (Divalent cations have large
effects on the hybridization rate, are often cofactors for nucleases, and are
avoided.) This term reflects the stabilization of duplexes by cations and is

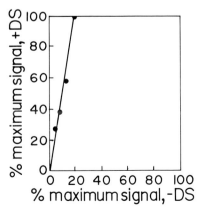

Figure 3-2 Dextran sulfate increases the
extent of hybridization of single-stranded
probes *in situ*. The extents of hybridization of
the histone probe to sections of 12-hour
embryos were compared in the presence (+DS)
and absence (−DS) of 10% dextran sulfate in
the hybridization solution. The points shown
for hybridization times ranging from 2 to 16
hours are reasonably fit by a straight line with
a slope of approximately 5, showing that fivefold higher signals are observed in the
presence of dextran sulfate at all hybridization
times. The probe concentration was 1.3
µg/ml.

probably accurate between 0.01 and 0.3M. Increasing M above 0.3 has progressively less effect on T_m (see Wetmur and Davidson, 1968).

L is the length of the duplexes formed. Note that for randomly sheared probes (nick translated DNAs or alkaline hydrolyzed RNAs) the length of duplex formed will always be shorter than the probe fragment length (Smith et al., 1975). Measurements by Cox et al. (1984) show that the T_m values of hybrids formed *in situ* are 5° C lower than those of hybrids formed in solution under identical conditions using 150- to 200-nt fragments. This probably reflects, at least in part, a shorter length for duplexes formed with target RNAs *in situ,* perhaps controlled by the distance between fixative cross-links. It is possible that the difference between solution and *in situ* T_m values may vary several degrees for tissues fixed by different protocols.

The %F term is the percentage of formamide. Formamide destabilizes nucleic acid duplexes and is used to lower the temperature required to achieve a given fidelity of hybridization. Lower temperatures reduce degradation of probe and loss of tissue from sections. The value of the constant (0.35° C × % formamide) holds only for RNA–RNA duplexes (Cox et al., 1984). The corresponding constant is 0.65 for DNA–DNA duplexes (McConaughy et al., 1969) and approximately 0.5 for DNA–RNA hybrids (Casey and Davidson, 1977). Note that the T_m values for RNA–RNA duplexes are also about 10° C higher than those of comparable DNA–DNA duplexes in the absence of formamide, which is incorporated in the first term of the preceding equation.

The %m term is the percentage of noncomplementary base pairs in the duplex formed. For detection of homologous sequences, $m = 0$.

The maximum hybridization rate in solution is observed at 25° C below the T_m value given by the preceding equation. The maximum rate *in situ* thus occurs at about 30° C below solution T_m. Because of the large number of variables involved (whose values and interdependencies are sometimes uncertain), it is not possible to predict the T_m more precisely than within a few degrees. However, hybridization rate exhibits a rather broad optimum, and is not much lower within 5° C of the T_m (Wetmur and Davidson, 1968; Brahic and Haase, 1978; Cox et al., 1984).

Hybridization at the temperature providing optimum rate will permit duplexes to form (although at reduced rate) with as much as 20% mismatch. If the probe sequence diverges from the target sequence (e.g., if the probe is from a different species), then the temperature for hybridization should be reduced accordingly. At $T_m - 25°$ C, most probes hybridize specifically to their target mRNAs. In some cases a probe may have sequence homology to other mRNAs, whose cross-reaction must be suppressed. This can be done, in theory, by hybridizing at a temperature closer to the T_m, but this also reduces the rate of hybridization for homologous duplexes. To discriminate between homologous and heterologous hybrids, we have found it much preferable to hybridize a set of slides at the normal temperature and then, after hybridization, wash different slides at progressively higher stringencies. An illustration of this approach can be found in Angerer et al. (1985b).

Posthybridization Washes

Removal of Mineral Oil

Wipe off as much excess as possible, being careful not to move the cover-slips, since the shear force may damage the sections. Place the slides in glass carriers and pass them through three changes of chloroform. If processing a large number of slides, you can reuse early washes, but the last wash should be done with clean chloroform to ensure that all oil is removed. Surface tension usually keeps the coverslips on during these washes, provided the slides are not too vigorously agitated.

Removal of Coverslips and Most of Probe

Drain the excess chloroform *briefly* and transfer the slides to $4 \times$ SSC at room temperature. Do not let the chloroform evaporate completely, as this may leave a residue of oil. Pass the slides through a total of three or four changes of $4 \times$ SSC at room temperature for a total of about 15 minutes to loosen the coverslips and wash off the bulk of unhybridized probe. This method is simple and avoids tissue loss.

RNase A Digestion

At a salt concentration of 20 μg/ml in 0.5 M NaCl, 10 mM Tris HCl, pH 8.0, 1 mM EDTA, for 30 minutes at 37° C, we see little, if any, effect of differences in RNase concentration between 20 and 50 μg/ml. This wash is economically done in small slide mailers, each holding five slides in 10 ml, which conserves enzyme. For detecting less abundant RNAs, it may be possible to reduce backgrounds further by including 1 unit RNase T1 per milliliter, because this enzyme hydrolyzes some sequences that are resistant to RNase A. Rinse all the glassware that comes in contact with RNase A in 0.1 NaOH to inactivate the enzyme. We keep separate the slide mailers used for RNase.

Final Washes

1. Wash in RNase A buffer for 30 minutes, at 37° C, in a staining dish or Coplin jar (about 10–15 ml/slide).
2. Wash in $2 \times$ SSC, for 30 minutes at room temperature. Place slides vertically in a test tube rack and immerse the rack in 4 liters of $2 \times$ SSC in a plastic rodent cage. The wash is stirred with a magnetic stirbar placed underneath the rack. The slides fit diagonally in the test tube rack, so that the buffer flushes their surfaces.
3. A wash in $0.1 \times$ SSC, 500 ml for 20 slides, at 50° C for 15 minutes achieves approximately the same stringency as the hybridization con-

ditions. Higher temperatures can be used to improve background without significant loss of signal (see below).

4. Wash in 4 liters of 0.1 × SSC for 30 minutes as in step 2, at room temperature.

Wash Procedure for ^{35}S or to Obtain Lower Backgrounds with ^{3}H

The following wash protocol has been required to achieve acceptable backgrounds with ^{35}S-labeled probes. It is also useful with ^{3}H-labeled probes when the target mRNAs are not very abundant.

1. After removal of coverslips and washes in 4 × SSC, as previously described, but with the addition of DTT to 10 mM, dehydrate the sections through graded ethanol concentrations containing 300 mM ammonium acetate, as described in the following on Dehydration.
2. Immerse the slides in hybridization buffer (0.3 M NaCl, 50% formamide, 20 mM Tris HCl, 1 mM EDTA, at pH 8.0) containing 10 mM DTT (10 ml/slide) and hold it for 10 minutes at 60 to 70° C. We have found that very stringent washes (only about 5° C below the *in situ* T_m) afford very low backgrounds with relatively little loss of signal.
3. Transfer to cold 2 × SSC and proceed as already described, beginning with the RNase digestion step. DTT is not included in these washes.

Dehydration

Dip the slides in glass carriers up and down for about 10 seconds each in 30%, 50%, 70%, 85%, and 95% EtOH, each containing 300 mM ammonium acetate (Brahic and Haase, 1978) as a precaution against denaturation of *in situ* hybrids. Then pass the slides through two changes of 99% EtOH and air dry them.

AUTORADIOGRAPHY

All steps are most easily carried out using the Duplex Super Safelight, equipped FDY (from VWR Scientific) filters in the top slots and FDW filters in the bottom slots. Our test showed that 3 hours of exposure to this light caused no detectable increase in emulsion background on slides. Alternatively, work can be done in absolute darkness with the use of a safelight equipped with a Wratten red filter No. 1.

Preparing Emulsion (Kodak NTB-2)

1. Melt emulsion for about 30 minutes by immersing the bottle in a 45° C water bath.

2. Dilute emulsion 1:1 with 600 m*M* ammonium acetate, also warmed to 45° C.
3. Aliquot in 10-ml batches in plastic scintillation vials. Double wrap the vials with aluminum foil.
4. Place the vials in the second container, which is light tight, and store it at 4° C.

Dipping Slides

1. Melt a vial of emulsion in a 45° C water bath (30–45 minutes). About 35 slides can be covered at least halfway with this volume, using the dipping chamber sold by Electron Microscopy Sciences.
2. Place the dipping chamber in the 45° C waterbath. Pour the emulsion in slowly to avoid bubbles.
3. Allow the emulsion to sit for 10 to 15 minutes to let bubbles rise to the top, and then dip some blank slides until surface bubbles have been removed.
4. Dip the slides by immersing them slowly twice, allowing about 2 seconds for each dip. This should be done as smoothly as possible, but trying to remove the slides very slowly from the emulsion leads to uneven coating.
5. Holding the slide vertically, blot the bottom edge on a paper towel for 1 to 2 seconds to remove excess emulsion. Place the slides vertically in a test tube rack to dry.
6. Transfer the rack containing the slides to an airtight chamber containing dampened paper towels and sealed with large sheets of parafilm 30 minutes after the last slide is dipped. Avoid plastic wraps that generate sparks. The high humidity removes latent grains from the emulsion.
7. Allow the slides to dry on the bench for about 45 minutes. The emulsion must dry slowly and completely; attempts to dry it too fast (by putting it under a vacuum, e.g.) lead to stretching artifacts, which result in grains forming along the edges of the tissue.
8. Transfer the slides to black plastic slide boxes, being careful to handle them only by the blank end and the opposite edge. It is important to dessicate the slides completely during autoradiography. The box containing a packet of dessicant is then sealed in a heat-sealable bag that also contains dessicant. Slides are exposed at 4° C in a larger light-tight box.

Developing

1. Remove the black boxes from the light-tight box and allow them to come to about 15° C while still sealed (about 15 minutes at room tem-

perature). This avoids condensation on the slides, which reduces the signals.

2. Carry out all the steps at a temperature of 15° C. Place the slides in a glass slide carrier and immediately develop the emulsion for 2.5 minutes in D-19 developer. Longer developing times and higher temperatures apparently preferentially produce grains in the emulsion background.
3. Stop the development with 2% acetic acid for 30 seconds.
4. Fix in Kodak fixer (not Rapid Fix) for 5 minutes.
5. Rinse in distilled water for 15 minutes.
6. Rinse in cold running tap water for 30 minutes.

Note. Developer, stop, and fixer should all be used at the same temperature; otherwise the emulsion may crack and flake.

Staining

1. Consult Rogers (1979) for a discussion on staining in conjunction with autoradiography. Many staining protocols (e.g., hematoxylin, eosin, Giemsa, methylene blue, methyl green-pyronin) are suitable, but procedures that use prolonged acid destaining bleach grains (Rogers, 1979; also our observations).
2. Mount the coverslips:
   a. After passing the slides through a graded ethanol series, dip them in several changes of fresh xylene.
   b. Remove a slide and drain the excess xylene by blotting the edge of the slide on a paper towel.
   c. To the moist slide, add 1 to 4 drops of Permount (depending on the coverslip size), add the coverslip, press it gently to force out any bubbles, and wipe the excess Permount from its edges.
   d. Allow the Permount to harden by placing the slide on a 40° C slide warmer for several hours.
   e. Clean the slides with xylene to remove the excess Permount, and with Formula 409 (or equivalent) to remove emulsion from the backside of the slide. Immersion oil can be removed with ethanol.

CONTROLS

Different experimental systems afford a variety of potential control experiments. In this we will describe several that are applicable in most cases.

Heterologous Probes

We strongly recommend the use of heterologous probes of the same specific activity, concentration, and fragment length to control for variations in background among cell types or tissues. Ideally, these are tested on sections adjacent to those hybridized with the experimental probes. Using moderately stringent wash procedures, nonspecific hybridization to mRNAs other than the desired target is not usually a significant source of background. Instead, background derives largely from binding to unidentified cellular constituents, and may therefore vary among cell types. Measurements of signal and nonspecific background using different posthybridization wash conditions allow maximization of signal-to-noise ratios.

mRNA Hybridization Efficiency

Does mRNA in all cells hybridize with equal efficiency? This question becomes significant in cases in which a tissue contains quite different cell types, which may respond differently to fixation and prehybridization treatments. There is no completely satisfactory solution to this potential problem, but several observations can help. In some cases, hybridization patterns are essentially "all or none," and hybridization specificity is demonstrated by hybridization of different probes to different subsets of cells. Use of a general probe, expected to hybridize to all cells at similar levels, can control for large differences in hybridization efficiency among cell types. Poly-U is probably one of the better choices, since all cells contain poly-A RNAs, although not necessarily at identical concentrations (e.g., see Lynn et al., 1983). In other cases, differences in signals of severalfold may be observed, and the question of relative hybridization efficiency becomes more significant. One possible control is to show that the ratio of grain densities over the different regions is the same for different fixation conditions, and extents of proteinase digestion—or reaches a constant value as proteinase concentration is increased.

Comparisons with External Data

In some cases probes may be available for RNAs whose concentrations in different stages of development or physiological conditions have been measured by other techniques, and measurement of *in situ* signals can be shown to accurately reflect such differences. Immunohistochemical data on localization of the protein product may also be suitable for comparison.

Demonstration That Signals Derive from Hybridization to RNA or DNA

Prehybridization treatments with RNase or DNase have been used as controls to identify the target nucleic acids as DNA or RNA. In most cases asymmetric transcription of cellular genes results in sense strand probes that are specific for DNA. This eliminates the need to remove target RNAs from sections, which introduces a variable, is difficult to achieve completely, and also requires additional controls to demonstrate that destruction of the probe by residual RNase is not responsible for reduced signals. With glutaraldehyde-fixed material, we have never detected signals from DNA even, for example, with sense probes complementary to 3600 kb of histone repeat DNA (DeLeon et al., 1983). In the absence of intentional denaturalization, as indicated previously, no step in the procedure is expected to denature DNA.

Plate 2 shows an example of the use of antisense and sense probes to detect human papilloma virus (HPV) DNA and RNA in a genital wart. Hybridization of the antisense strand to tissue in which HPV DNA has not been denatured to allow hybridization affords specific detection of HPV mRNA (Plate 2a); hybridization with the sense strand to tissue in which HPV DNA has been denatured provides specific detection of the DNA (Plate 2b); hybridization of the sense strand to tissue in which the DNA has not been denatured (Plate 2c) should not yield any specific signal. Although we have not detected nuclear DNA sequences in glutaraldehyde-fixed tissue, we do see some specific hybridization of sense strands to nuclei in formalin-fixed tissue, even when this has not been intentionally denatured. Since formaldehyde slowly denatures DNA at neutral pH, we believe these signals represent partial denaturation of DNA during the fixation procedure. In cases in which tissue contains a significant amount of target DNA sequence (e.g., in viral infections) a nonhomologous probe should be used to measure nonspecific background binding.

SENSITIVITY

General Comments

The sensitivity of *in situ* hybridization depends directly on the signal-to-noise ratio that can be achieved. The signal depends on (1) the fraction of target RNA retained in the tissue, (2) the fraction of retained RNA that can be made accessible to probe, (3) the mass of probe that hybridizes in relation to mass of target RNA, and (4) the visible signal that can be produced per nucleotide of hybridized probe. The noise depends on the extent to which nonspecific sticking of probe can be prevented and on emulsion background. Note that background is always higher over tissue than in the surrounding emulsion.

The first three of these factors depend on optimization of fixation and prehybridization treatments. Results from several laboratories using somewhat different protocols indicate that most of the target RNA can be retained and hybridized with single-stranded probes (Brahic and Haase, 1978; Cox et al., 1984). Using radioactively labeled probes and autoradiographic detection of signals, it is possible to measure directly the signal achieved for a certain mRNA concentration (e.g., see Cox et al., 1984). Similar calculations show that at saturation ^{3}H- and ^{35}S-labeled probes complementary to 500 nucleotides (nt) of an mRNA (at specific activities of 2.5×10^8 and 2.5 and 10^9 dpm/μg, respectively) would yield about 0.036 and 1.8 grains/ 100 μm^2/day for a message density of 1 copy/cell (a fixed cell is about 25 μm^3 in our system). A grain density of 10 grains/100 μm^2 is readily detectable in the absence of nonspecific binding. This emphasizes the fact that the current limitations on sensitivity (signal-to-noise) is the level of nonspecific binding of probe.

Choice of Detection Mechanism

Radioisotopes

1. ^{3}H. Except for very abundant targets, we synthesize probes using UTP and CTP at the specific activities (25–60 Ci/mmol) supplied by the manufacturers. It is also possible to use labeled GTP, although it is less stable than the other precursors and available only at lower specific activity. Several suppliers currently offer rXTPs that have been assayed and shown to be free of an unidentified inhibitor, which in the past sometimes severely suppressed SP6 polymerase activity. Currently available precursors provide probe-specific activities equal to that of the precursor nucleotide pool, 1 to 20×10^8 dpm/μg. ^{3}H permits the best resolution of any isotope, but the lowest autoradiographic efficiency (approximately 0.02 grains/disintegration in 5-μm sections; Ada et al., 1966; Pelc and Welton, 1967). Other advantages include stability and a long half-life.

2. ^{35}S. SP6 polymerase will incorporate ^{35}S-substituted XTPs, but at about a twofold reduced rate. Therefore, ^{35}S-substituted precursors should be cut with cold S-substituted nucleotide. Using one ^{35}S-nucleotide triphosphate (UPT or ATP), it is possible to obtain specific activities 10 times higher than with ^{3}H precursors, and the autoradiographic efficiency is about 5 times higher. The resolution is adequate for most applications, except possibly those requiring discrimination between nucleus and cytoplasm in small cells. Thus, ^{35}S provides 50-fold higher signals. The extent to which this isotope affords higher sensitivity (as opposed to shorter exposure times without better signal/noise), then, depends on the extent to which background can be suppressed. In recent experiments the previously outlined modifications have provided

^3H ^{35}S

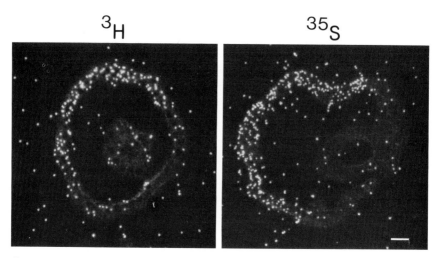

Figure 3-3 Sensitivity of probes labeled with ^{35}S. A probe complementary to 131 nt of the 3' end of a sea urchin cytoplasmic actin mRNA (CyIIIa; Lee et al., 1984) was hybridized to sections of sea urchin blastulas: (a) ^{3}H-labeled; specific activity 1.8×10^8 dpm/μg; autoradiographic exposure 60 days; (b) ^{35}S-labeled, specific activity 5×10^8 dpm/μg; autoradiographic exposure 6 days. The probe concentration was 0.07 μg/ml in both (a) and (b), which is calculated to achieve saturation of target RNAs. From available data (Lee et al., 1986) and observed grain densities, we estimate that there are approximately 125 molecules of this mRNA/cell in the region of the gastrula that contains this message.

low backgrounds, and a consequent significant increase in signal-to-noise. Figure 3-3 compares the same probe labeled with ^{35}S (5×10^8 dpm/μg) and with ^{3}H (1.8×10^8 dpm/μg), showing a higher signal-to-noise for ^{35}S at a 10-fold shorter exposure time.

Note that it is unwise to use high-specific-activity ^{35}S-labeled probes at concentrations far below those required to saturate target RNAs. In these cases, better resolution and similar signal levels can be achieved at lower cost by using saturating concentrations of lower-specific-activity ^{3}H-labeled probes.

3. ^{32}P. Probes of very high specific activity can be obtained, and such probes have been quite successfully used in conjunction with x-ray film autoradiography to identify organs, or regions of organs, containing individual mRNAs, as is amply illustrated in other chapters in this volume. The resolution is not suitable for finer-scale localization of labeled cells. Direct comparison in our laboratory indicates that the autoradiographic efficiency (grains per disintegration) using our standard procedure is not much higher than that obtained with ^{3}H.

4. ^{125}I. This isotope offers the potential advantage of double detection by x-ray film at the macroscopic level, and liquid emulsion at the light level. ^{125}I-labeled CTP is not readily available commercially, and the large amounts of this isotope required for do-it-yourself synthesis of

^{125}I-CTP constitute a significant health hazard. Using CTP labeled to approximately 2000 Ci/mmol, we found that iodinated precursor could be incorporated by SP6 RNA polymerase, but the product was extremely unstable, and the specific activity of the transcripts decreased rapidly over a period of 1 or 2 days (our unpublished observations).

Nonradioactive Detection Methods

These are based on synthesis of probes containing a tag, for example, biotin or other small molecules, and detection of hybrids with either antibodies or streptavidin, which in turn can be visualized fluorometrically or cytochemically (Brigati et al., 1984). The advantages of this approach are excellent resolution and speed of detection; the disadvantages are that the signal is difficult to quantitate and the sensitivity is relatively low. Based on the fact that shorter probes give higher signals, we suggest that the low sensitivity results from poor penetration of large reporter molecules to hybridized probe.

EPILOGUE

It appears that the time of technical trials and tribulations will soon be past. *In situ* hybridization methods are, or soon will be, available for a variety of tissues, and we can all pursue the interesting biological questions. Some technical advances are still to be hoped for, for example, in facilitating data analysis, and in use of nonisotopic detection methods. We emphasize that *in situ* hybridization is based on rational principles, that major parameters have been characterized, and that variations of these parameters have predictable effects. While we can enjoy the successes of this technique, it is prudent to remain aware of its limitations and potential pitfalls. Application of the technique without understanding favors the pitfalls.

We appreciate comments on the successes and failures of our particular protocol and welcome suggestions for its improvement.

Acknowledgments

The experiments of Figures 3-1 and 3-2 were carried out by Robert West in the laboratory of RCA and LMA. Research in the laboratory of RCA and LMA is supported by National Institute of Health (NIH) GM 25553. RCA is the recipient of a Research Career Development Award from the NIH. The initial application of these techniques to formalin-fixed clinical samples was supported by a Biomedical Research Support Grant (2-S07-RR05403-24) to MHS.

Sources of materials

The following list includes the less common equipment and chemicals used in our protocol. This is designed as a convenience and to indicate the general grade of chemicals used, and is not necessarily an endorsement of any particular source.

Equipment

Heat block	VWR No. 13259-005 and No. 13259-130
Paraffin oven	Boekel No. 131700
Super Duplex Safelight	VWR No. TM72882-10 with FDY top and FDW bottom filters

Enzymes

Proteinase K	EM Laboratories No. 24568-10
RNase inhibitor	Promega Biotec No. P211
DNase (DPRF)	Worthington No. LS0006333
RNase A	Worthington No. LS-05650

Chemicals

Diethylpyrocarbonate	Sigma No.D-5758
Glutaraldehyde	Sigma No.G5882 (vial used once and discarded)
Poly-L-lysine	Sigma No.P1399
Paraplast	VWR No.15159-409
Tris, ultra pure	Schwartz/Mann No.819620
Triethanolamine	VWR JT-9467-1
Acetic anhydride	VWR JT-0018-1
Dithiothreitol	BioRad No.161-0611
ATP	Pharmacia Molecular Biochemicals, No.27-1006-01
CTP	Pharmacia Molecular Biochemicals, No.27-1200-02
UTP	Pharmacia Molecular Biochemicals, No.28-0700-01
GTP	Pharmacia Molecular Biochemicals, No.27-2000-02
Yeast tRNA	BRL No.5401SB
Spermidine	Sigma No.52501
Phenol	IBI No.05164
Formamide	IBI No.72024 (recrystallized 2× and deionized for hybridization; deionized only for washes)
BSA	BRL No.5561UA
Ficoll	Sigma No.F-4375
Polyvinyl pyrrolidone	Sigma No.PVP-360
Dextran sulfate	Sigma No.D-6001
Emulsion, NTB-2	Eastman Kodak No.1654433

All other chemicals are reagent grade. All solutions are prepared with water that is charcoal filtered, deionized, and stirred with 0.1% diethylpyrocarbonate (DEP) overnight, and autoclaved. Stock solutions of chemicals are millipore filtered and DEP treated (unless they are reactive with DEP).

Accessories

Embedding molds	VWR No.15160-157
BEEM capsules	Polysciences No.0224
Slide mailers	Thomas No.6707-M25
Test tube rack	VWR No.60939-005
Staining dish and slide carrier	Wheaton, American Scientific Products, No.900300
Coplin jar	Wheaton, VWR No. 2546-000
Dipping chamber	Electron Microscopy Sciences, No.07051

REFERENCES

Ada, G.L., Humphrey, J.H., Askonas, B.A., McDevitt, H.D., and Nossal, G.V. Correlation of grain counts with radioactivity (^{125}I and ^{3}H) in autoradiography. *Exp. Cell Res.* 41:557–572, 1966.

Angerer, L.M. and Angerer, R.C. Detection of poly-A RNA in sea urchin eggs and embryos by quantitative *in situ* hybridization. *Nucleic Acids. Res.* 9:2819–2840, 1981.

Angerer, R.C., Cox, K.H., and Angerer, L.M. *In situ* hybridization to cellular RNAs. *Genet. Eng.* 7:43–65, 1985a.

Angerer, L., DeLeon, D., Cox, K., Maxson, R., Kedes, L., Kaumeyer, J., Weinberg, E., and Angerer, R. Simultaneous expression of early and late histone messenger RNAs in individual cells during development of the sea urchin embryo. *Dev. Biol.* 112:157–166, 1985b.

Bodkin, D.K. and Knudson, D.L. Assessment of sequence relatedness of double-stranded RNA genes by RNA–RNA blot hybridization. *J. Virol. Methods* 10:45, 1985.

Brahic, M. and Haase, A.T. Detection of viral sequences of low reiteration frequency by *in situ* hybridization. *Proc. Natl. Acad. Sci. USA* 75:6125–6129, 1978.

Brigati, D.J., Myerson, D., Leary, J.J., Spalholz, B., Travis, S.Z., Fong, C.K.Y., Hsiung, G.D., and Ward, D.C. Detection of viral genomes in cultured cells and paraffin-embedded tissue sections using biotin-labeled hybridization probes. *Virology* 126:32–50, 1983.

Bruskin, A., Tyner, A.L., Wells, D.E., Showman, R.M., and Klein, W.H. Accumulation in embryogenesis of five mRNAs enriched in the ectoderm of the sea urchin pluteus. *Dev. Biol.* 87:308–318, 1982.

Butler, E. and Chamberlin, M.J. Bacteriophage SP6-specific RNA polymerase. *J. Biol. Chem.* 257:5772–5778, 1982.

Casey, J. and Davidson, N. Rates of formation and thermal stabilities of RNA:DNA and DNA:DNA duplexes at high concentrations of formamide. *Nucleic Acids Res.* 4:1539–1552, 1977.

Cox, K.H., DeLeon, D.V., Angerer, L.M., and Angerer, R.C. Detection of mRNAs in sea urchin embryos by *in situ* hybridization using asymmetric RNA probes. *Dev. Biol.* 101:485–502, 1984.

DeLeon, D.V., Cox, K.H., Angerer, L.M., and Angerer, R.C. Most early-variant histone mRNA is contained in the pronucleus of sea urchin eggs. *Dev. Biol.* 100:197–206, 1983.

Gall, J.G. and Pardue, M.L. Nucleic acid hybridization in cytological preparations. *Methods Enzymol.* 21:470–480, 1971.

Gall, J.G., Stephenson, E.C., Erba, H.P., Diaz, M.V., and Barsacchi-Pilone, G. Histone genes are located at the sphere loci of newt lampbrush chromosomes. *Chromosoma* 84:159–171, 1981.

Haase, A.T., Stowring, J.D., Harris, B., Traynor, B., Ventura, P., Peluso, R., and Brahic, M. Visna DNA synthesis and the tempo of infection *in vitro*. *Virology* 119:399–410, 1982.

Hayashi, S., Gillam, I.C., Delaney, A.D., and Tener, G.M. Acetylation of chromosome squashes of *Drosophila melanogaster* decreases the background in autoradiographs from hybridization with [^{125}I]-labeled RNA. *J. Histochem. Cytochem.* 26, 677–679, 1978.

Huang, W.M., Gibson, S.J., Facer, P., Gu, J., and Polak, J.M. Improved section adhesion for immunocytochemistry using high molecular weight polymers of L-lysine as a slide coating. *Histochemistry* 77:275–279, 1983.

Krieg, P.A., Rebagliati, M.R., Green, M.R., and Melton, D.A. Synthesis of hybridization probes and RNA substrates with SP6 polymerase. *Genet. Eng.* 7:165–184, 1985.

Krieg, P.A. and Melton, D.A. In *Promega Notes,* No. 1, 1985.

Lee, J.J., Shott, R.J., Rose III, S.J., Thomas, T.L., Britten, R.J., and Davidson, E.H. Sea urchin actin gene subtypes: Gene number, linkage and evolution. *J. Mol. Biol.* 172:149–176, 1984.

Lee, J.J., Calzone, F.C., Britten, R.J., Angerer, R.C., and Davidson, E.H. Activation of sea urchin actin genes during embryogenesis—measurement of transcript accumulation from 5 different genes in Strongylocentrotus-purpuratus. *J. Mol. Biol.* 188:173–183, 1986.

Lynn, D.A., Angerer, L.M., Bruskin, A.M., Klein, W.H., and Angerer, R.C. Localization of a family of mRNAs in a single cell type and its precursors in sea urchin embryos. *Proc. Natl. Acad. Sci. USA* 80:2656–2660, 1983.

Maniatis, T., Fritsch, E.F., Sambrook, J. *Molecular Cloning: A Laboratory Manual,* Cold Spring Harbor Laboratory, 1982.

McAllister, L.B., Scheeler, R.H., Kandel, E.R., and Axel, R. *In situ* hybridization to study the origin and fate of identified neurons. *Science* 222:800–808, 1983.

McClay, D.R., Wessel, G.M., and Marchase, R.B. Intercellular recognition: Quantitation of initial binding events. *Proc. Natl. Acad. Sci. USA* 78:4975–4979, 1981.

McConaughy, B.L., Laird, C.D., and McCarthy, B.J. Nucleic acid reassociation in formamide. *Biochemistry* 8:3289–3295, 1969.

Melton, D., Krieg, P., Rebagliati, M., Maniatis, T., Zinn, K., and Green, M.R. Efficient *in vitro* synthesis of biologically active RNA and RNA hybridization probes from plasmids containing a bacteriophage SP6 promoter. *Nucleic Acids Res.* 12:7035–7056, 1984.

Millonig, G. and Marinozzi, V. Fixation and embedding in electron microscopy. In *Advances in Optical and Electron Microscopy,* Volume 2. R. Barer and V.E. Cosslett (Eds.). London and New York: Academic Press, 1968, pp. 251–336.

Pelc, S.R. and Welton, M.G.E. Quantitative evaluation of tritium in autoradiography and biochemistry. *Nature* (London) 216:925–927, 1967.

Rogers, A.W. *Techniques in Autoradiography.* New York: Elsevier/North Holland Biomedical Press, 1979.

Smith, M.J., Britten, R.J., and Davidson, E.H. Studies on nucleic acid reassociation kinetics: Reactivity of single-stranded tails in DNA-DNA renaturation. *Proc. Natl. Acad. Sci. USA* 72:4805–4809, 1975.

Wetmur, J.G. and Davidson, N. Kinetics of renaturation of DNA. *J. Mol. Biol.* 31:349–370, 1968.

4

Toward a Rapid and Sensitive *In Situ* Hybridization Methodology Using Isotopic and Nonisotopic Probes

Robert H. Singer, Jeanne Bentley Lawrence, and Roya N. Rashtchian

In situ hybridization has become an important molecular tool since the development of recombinant DNA technology. Because of the ability to obtain purified sequences of DNA, the hybridization of these sequences to nucleic acids *in situ* reveals information concerning the expression of particular genes or the presence of infectious agents. Hybridization to chromosomes reveals the location of particular genes of interest.

Conventional approaches using molecular biology to study gene expression rely on isolation of nucleic acid sequences from a mass of tissue or cells, thereby averaging the information from each individual cell with the total contribution of all cells. Because most biological systems are very heterogeneous in cell type and in the timing of molecular events, *in situ* hybridization provides the only method to correlate molecular information with biochemical or morphological markers on individual cells. Furthermore, because *in situ* hybridization can be used on very small amounts of tissue, such as an early embryo, it makes possible the detailed examination of the complex series of molecular events occurring during development. Additionally, *in situ* hybridization allows detection and localization of sequences diagnostic of pathogenesis. Hence viral infections, oncogenes, parasitic, or other etiological agents of disease can be detected in a cell sample or biopsy.

Because of the importance of this technique for these and other endeavors, we undertook to develop a methodology that was sensitive, reproducible, and convenient. Most previous work using *in situ* hybridization has not focused on the methodology per se (see, however, Cox et al., 1984; Haase et al., 1985). Our intention was to optimize *in situ* hybridization methodology with respect to morphology, sensitivity, and retention of nucleic acid sequences. In approaching this goal we have not only characterized parameters affecting hybridization of probe to cellular nucleic acids, but have also evaluated different isotopic and nonisotopic methods for detecting that probe once it is hybridized within the cell.

We have undertaken an analytical procedure that has allowed us to evaluate hybridization rapidly and quantitatively, based on the use of a ^{32}P-labeled probe. The cells, tissues, or chromosomes are hybridized to this probe under various experimental conditions, and then the amount of probe hybridized per sample is quantitated in a scintillation counter by Cerenkov radiation. After quantitation in a physiologically compatible solution, the samples can be microscopically examined for preservation of morphology. Additionally, cells can be labeled with tritiated uridine and the retention of RNA directly measured by scintillation counting on the same samples for which hybridization of probe is also evaluated. This allows a rapid and quantitative assessment of parameters that affect hybridization, to optimize or eliminate them. After the development of a convenient methodology for isotopically labeled probes, it was then possible to undertake a similar evaluation of probes labeled simultaneously with ^{32}P and a biotin analog to assess the *in situ* hybridization of nonisotopically modified probes. The detection of biotin-labeled probes can be further assessed using ^{125}I-labeled avidin to evaluate and reduce background contributed by the detector.

For the development of a protocol for cells and tissues we used a well-defined model system that is also the subject of our research interest. For the bulk of this work we used primary chicken fibroblasts and myoblasts. The myoblasts differentiate into multinucleated myotubes, which then synthesize large amounts of actin and other muscle-specific proteins. The probe used for these studies was a full-length chicken beta-actin cDNA probe obtained from Don Cleveland and cloned into pBR322. The actin clone was used because actin is expressed in all cells in high abundance (2000 copies per cell) and is particularly expressed in differentiated muscle (10,000 copies per nucleus). For optimizing hybridization to human chromosomes, the entire human genome was used as a probe so that the extent of hybridization could be quantitated rapidly in the scintillation counter.

Some of the results of this approach, discussed in the following material, have been previously published (Lawrence and Singer, 1985). These results are described briefly and additional results on tissues, nonisotopic probes, and chromosomes are considered as well. For application to neurobiology, we have used a POMC probe on tissue sections of rat pituitary.

RESULTS

Fixation

Any procedure involving *in situ* hybridization begins with the fixation of the tissue, cell, or chromosomes. The fixation must preserve the material in a morphologically stable state so that the nucleic acids are retained in position without diffusion, particularly under the rigorous conditions used for hybridization. Extensive cross-linking of proteins, however, renders the cytoplasm relatively impermeable to all but the smallest probes. Our protocol is based on the need to maintain morphological information as well as to obtain the most efficient hybridization. In optimizing this protocol, we monitored cellular RNA retention and quantitated the extent of hybridization in the same samples. After analyzing many fixatives in this way, we concluded that cell fixatives such as paraformaldehyde and glutaraldehyde, which are known to cross-link proteins, are acceptable for RNA retention (see Table 4-1). Of these two fixatives, paraformaldehyde (a solid polymer of formaldehyde) does not cross-link the cell so extensively that it prevents penetration of large probes (see the section Probe Size). We find bottled formalin solutions to be unpredictable in RNA retention because of degradation; therefore, we make fresh solutions of paraformaldehyde every 2 weeks. Precipitating (non-cross-linking) fixatives are insufficient for cellular RNA retention under the conditions used (e.g., Carnoy's or ethanol/acetic acid), but some of the fixatives commonly in use in clinical pathology laboratories appear adequate. A further investigation on glutaraldehyde was warranted because it is a commonly used fixative, preserves cellular morphology better, and is at least as good as paraformaldehyde for preserving RNA. Table 4-2 shows that a 1% solution of glutaraldehyde at $0°$ C (see Angerer and Angerer, 1981) can be used in conjunction with significantly smaller probe sizes (less than 250 nucleotides) and proteinase digestion to yield results comparable to those of paraformaldehyde (see the sections on Probe Size and Cell Pretreatments). The ability to use a variety of probe sizes with paraformaldehyde without further treatments as well as its lower intrinsic autofluorescence (see Biotinylated Probes section) make glutaraldehyde a superior fixative for our experimental purposes.

We found the fixation protocols used with cultured cells to be transferable

Table 4-1 Signal, noise, and RNA retention with different fixatives[a]

	Signal	Noise	RNA Retention
Paraformaldehyde (4%)	100	100	100
Glutaraldehyde (4%)	44	130	120
Buffered formalin (4%)	45	100	46
Zenker's	100	600	110
Bouin's	60	100	77
Osmium tetroxide (1%)	45	50	61
Acetic acid/ethanol (3:1)	32	100	38
Carnoy's	40	100	33

[a]Expressed as percentages of paraformaldehyde within each column.

Note: The signal obtained with paraformaldehyde using a ^{32}P-labeled actin probe was compared with the signal obtained with different fixations of the same cell preparations. All fixation was for 15 minutes at room temperature. The Cerenkov cpm obtained with the fixatives was divided by the cpm in paraformaldehyde and multiplied by 100 to obtain the percentage of hybridization relative to that of paraformaldehyde. The same was done for noise, using only pBR322 as probe. RNA retention was by scintillation counting of ^{3}H-uridine label within the cell after hybridization and the cpm was treated as with ^{32}P counts. Probe size was 400 to 600 nucleotides.

Table 4-2 Effect of various fixation times and proteinase treatments on
glutaraldehyde fixation at 0° C[a]

	Signal	Noise	RNA retention
15' Fixation, 4%,	44	120	120
With 10' proteinase K	82	100	100
With 30' proteinase K	86	100	90
15' Fixation 1%,	69	100	120
With 10' proteinase K	75	100	120
With 30' proteinase K	100	86	90
5' Fixation, 4%,	40	90	120
With 10' proteinase K	51	100	90
With 30' proteinase K	63	72	100
60' Fixation, 1%,	45	97	120
With 10' proteinase K	77	100	110
With 30' proteinase K	84	120	100

[a]Expressed as percentages of paraformaldehyde (4%) fixation at room temperature within each column (see Table 4–1). Proteinase K concentration at 5µg/ml—average of duplicate samples. Probe sizes less than 200 nucleotides were used.

to tissue sections. Particular technical problems with tissue sections are the control of fixation time and the adherence of the section to a coverslip or slide during the hybridization protocol. After experimenting with many slide preparation techniques such as subbing with potassium chromate, albumin, gelatin, poly-L-lysine, and others, we found that the best results are obtained when the unfixed tissue is frozen-sectioned and placed on glass slides or coverslips coated with Histostik (Accurate Scientific, New Jersey). The tissue was frozen in an isopentane solution cooled by liquid nitrogen and then stored in liquid nitrogen. The tissue was sectioned at −15° C on a cryomicrotome at 5 to 8 μm. Sections were picked up onto a Histostik-coated slide or coverslip and allowed to sit on the glass for 5 minutes at room temperature. They were then fixed for 15 minutes in paraformaldehyde (4%) and processed through the *in situ* hybridization protocol used for cultured cells.

For chromosomes, a completely different fixation protocol is required because RNA retention is not desired. The standard protocol (see Kozak et al., 1977) uses ethanol/acetic acid (3:1) to prepare chromosomes on slides. We have tested the effect of further fixation after denaturation using glutaraldehyde, formaldehyde, or osmium tetroxide, but have found no increase in signal with these treatments. With regard to the detection of viruses, the double-stranded DNA viruses may require a protocol more like that used for chromosomes (see, e.g., Haase et al., 1982). We have successfully detected double-stranded DNA viruses with the formaldehyde protocol (see also Brigati et al., 1983) but have not compared this method directly with acetic acid fixation. A direct comparison of formaldehyde fixation with acetic acid fixation using a single-stranded RNA retrovirus (HTLVIII, in which most viral sequences are in the form of RNA transcripts) as a model showed formaldehyde to be substantially better as a fixative.

After fixation, the cells or tissues can be stored indefinitely in 70% ethanol at 4° C, but chromosomes are dried and stored in the freezer. This allows "stockpiling" of cells of various types to be tested simultaneously with the same probe. The ethanol also solubilizes lipids in the cell membrane; however, this step is not obligatory because the hybridization solution appears to serve the same function. The cells can also be stored in physiological solution for 2 weeks at 4° C before they are hybridized, with results equivalent to those obtained with storage in ethanol.

Cell Pretreatments

Many protocols using *in situ* hybridization subject the cells to a variety of treatments in an attempt to increase hybridization by rendering target sequences more accessible to the probe. Most of these methods are directed toward permeabilization of the fixed cellular protein matrix. These include the use of protease, acid, detergents, or heat denaturation. Other pretreatments are used to reduce the background by inhibiting nonspecific binding of nucleic acids; examples are prehybridization with nonspecific DNA (e.g., salmon sperm) or RNA (e.g., tRNA), acetic anhydride, or Denhardt's solution.

We have analyzed a constellation of factors used in various *in situ* hybridization protocols and found that, after a short paraformaldehyde fixation, most cellular pretreatments are either unnecessary or detrimental to RNA retention and, consequently, to hybridization. A case in point, presented in Figure 4-1, is the use of protease (either pronase or proteinase K), in which case RNA loss is significant even with low concentrations of the enzyme. This is consistent with our hypothesis that paraformaldehyde cross-links the cell cytosol just enough to hold the RNA in the cell, but still leaves the RNA available for hybridization. Any additional digestion of the cell releases this RNA.

A different picture emerges with glutaraldehyde, with which probe penetration is evidently the limiting factor and additional permeabilization by proteolytic digestion is beneficial for the signal, as shown in Table 4-2. A direct comparison of glutaraldehyde with formaldehyde, using probe sizes of 100 to 200 nucleotides, shows both fixatives to be approximately the same, although proteinase digestion is still required for glutaraldehyde. Because we have emphasized the development of a simple protocol that is gentle to cell morphology, these results reinforce the choice of paraformaldehyde since this fixation obviates the need for further treatment of the cell, which can only be deleterious to morphological detail. Other treatments, such as acid or acetic anhydride, appear to make no significant difference, and we have eliminated them from our standard protocol. For specific experiments requiring use of very large probe fragments, however, cell pretreatment with acetic anhydride is valuable for reducing nonspecific interactions (see following).

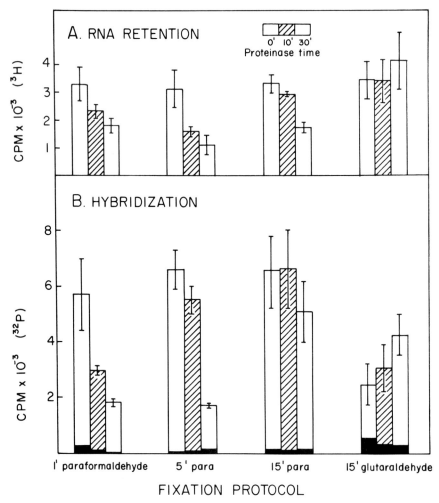

Figure 4-1 Effect of proteinase digestion on hybridization and RNA retention. Cells were labeled for 3 hours in ^{3}H uridine (10 μCi per 5 ml of medium), fixed as described, and then hybridized overnight with the actin probe (50,000 cpm/ng). The samples were counted by scintillation using toluene-liquifluor. Windows on the counter were set to read both ^{3}H and ^{32}P from the same samples. The black bars in B represent hybridization of the control pBR322 probe. Prior to hybridization, the samples were incubated for varying times in 5 μg/ml proteinase K in PBS plus MgCl$_2$. The results presented are the average of two experiments each of which used duplicate samples.

Probe Size

The probe must diffuse into the cell to hybridize to the appropriate sequences, and the size of the fragments after nick translation will limit this diffusion. To explore the influence of probe size on signal, the size of the fragments was varied by increasing the amount of DNase present in the nick-translation reaction. Paraformaldehyde-fixed cells tolerated a broad range of probe sizes without loss of signal (Figure 4-2). In contrast, we find that

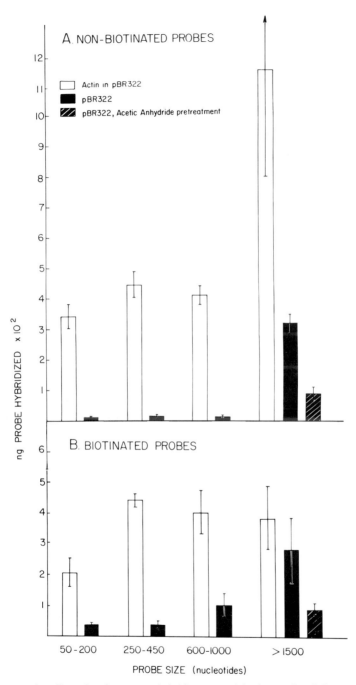

Figure 4-2 The effect of probe size on hybridization and background with biotinated and nonbiotinated probes. The cells were hybridized to probes of indicated sizes for 4 hours. Biotinated probes were labeled with biotin-11-dUTP and ^{32}P dCTP. Cells treated with acetic anhydride (0.25% in triethanolamine buffer pH 8.0 at room temperature) were also hybridized with these probes. The results are from two experiments that used duplicate samples. The bars indicate standard deviations.

glutaraldehyde-fixed cells require small fragments (below 250 nucleotides), as has been reported previously (Angerer and Angerer, 1981). This is consistent with the hypothesis that cells or tissues fixed with paraformaldehyde are less cross-linked and more permeable; thus, probe size is not as critical a factor. However, when biotinylated probes are used, when the thymidine in the nick-translation reaction is completely replaced by biotinylated dUTP (11 atom linker arm), the probe size must remain relatively small to avoid nonspecific sticking (Figure 4-2). This may be because the presence of the highly charged biotin moiety makes the DNA more rigid, and less able to penetrate or be rinsed away from the matrix of the cell.

We have found that removal of the insert (in this case the 2 kb of actin) from pBR322 did not improve hybridization. In fact, not only is it more convenient simply to nick-translate the insert in the vector but, as can be seen in the case that follows, it improves the signal. Enhancement of signal is seen when very large probes (1000 nucleotides or more) are used for *in situ* hybridization. This raises the possibility that these probes are capable of forming networks in which the vector contributes to the signal. If the recombinant plasmid is 6500 nucleotides (vector plus insert), then a probe size of 1500 nucleotides represents about four nicks by DNAse per recombinant molecule and, on the average, half of those would contain actin sequences contiguous with pBR322 sequences. Therefore, these vector sequences would provide "tails" on which to build further hybrids. Hence, a network of probe molecules would form a lattice structure on a single messenger RNA.

If this hypothesis is correct, networking could be prevented in two ways. The first is by using probes smaller than the insert (in this case, 450 nucleotides). This will reduce the probability of "junction pieces" containing both actin and pBR322; hence, the hybridization measured will represent only the probe to the target. The second way of preventing networking is to cut the insert at the point where it joins the plasmid (in this case, with the restriction endonuclease, Pst-1), thereby eliminating the continuity of actin and pBR322 sequences and hence the "junction pieces" on which the networks form. The results shown in Table 4-3 indicate that the data bear out this prediction. The increased signal obtained with large probe fragments, generated by nick translation of intact plasmid, is not observed with small probe fragments or with large probe fragments generated from plasmid

Table 4-3 Effect of Pst 1 digestion on hybridization of actin probe

Actin probe	Approximate fragment size	Nanograms of probe hybridized (range)
Uncut	450	0.032 (0.028–0.039)
Pst-1 cut	450	0.034 (0.028–0.045)
Uncut	1500	0.302 (0.101–1.05)
Pst-1 cut	1500	0.045 (0.03–0.07)

previously cut with endonuclease, so that the actin and pBR322 sequences are no longer contiguous. This indicates that networking is probably occurring in the large probes, and it allows a mechanism for signal enhancement at the hybridization level.

Because the signal is amplified by a factor of 10, on the average, and occasionally by as much as 25, the vector can be used to increase the detection of lower abundance messages. However, this approach must be used in conjunction with techniques to lower the nonspecific background; these include treatment with acetic anhydride or use of equivalently large nonspecific competitive inhibitors such as mildly sheared salmon sperm DNA. Since the use of large probes gives variable results (see the range in Table 4-3), they should not be used for quantitation that requires reproducibility. However, in comparing small probes generated either from the intact recombinant plasmid or the insert disconnected from the vector, there is no reproducible difference, indicating that networking is not contributing significantly to signal with probes of 450 nucleotides or fewer. Therefore, these probe sizes can be used quantitatively. For this reason, the quantitative analysis of *in situ* hybridization described in the next section has been done with probe sizes for which networking will not contribute to signal.

QUANTITATIVE CONSIDERATIONS

In situ hybridization has major advantages over solution or blot hybridization for quantitation of messages per cell. First, the approach is rapid and convenient and can be done with orders of magnitude fewer cells as starting material. Most important, because the cells are not disrupted, individual cells can be viewed with the appropriate detector (see the section on Detection) for information regarding gene expression within a heterogeneous cell population. An additional advantage is that, because the messenger RNA is effectively immobilized on a solid matrix (as if it were nitrocellulose bound), labeled DNA excess experiments can assess the copy number of that sequence directly, since unhybridized DNA is simply washed away. When the probe is small enough that networking does not occur, the hybridized labeled DNA corresponds to the number of target messages detected in the cell.

There are two differences between *in situ* hybridization and hybridization in solution or to blots where the nucleic acids have been extracted: First, the probe must penetrate the cell matrix to get to the target, and second, the target may be partially bound with proteins. In the first case, if the time necessary to penetrate the cell is significant in relation to the time double-stranded probes reanneal, the amount of hybridization may be prematurely truncated. In the second case, some fraction of target nucleic acids may not be available to hybridize. Messages are known to be bound to various proteins in addition to ribosomes, and these may inhibit hybridization *in situ* (see, e.g., Vincent et al., 1980). It was important, therefore, to test the efficiency of *in situ* hybridization and compare it to solution hybridization.

We used the developing muscle system to quantitate the detection of actin target sequences because this message changes over time and over an order of magnitude in a predictable manner. Schwartz and Rothblum (1981) have accurately measured these sequences using solution hybridization. Our first test of the quantifiability of the *in situ* system was to increase the DNA probe concentration to a point where the cellular targets would be saturated. At this point, we could determine the copy number of messenger RNAs available for hybridization after counting the radioactivity of the probe hybridized to a coverslip, converting these data to nanograms of DNA bound, and then dividing the result by the number of cells. The molecular weight of the actin probe allowed us to determine that each messenger should hybridize to 1.1×10^{-9} ng of single-stranded DNA, so that the number of molecules hybridized per cell could be determined. Furthermore, this approach could be applied to cell populations expressing varying amounts of actin messenger RNA. If the technique yields quantitative data, it will distinguish differences in gene expression between populations.

The results in Figure 4-3 indicate that the method can be used for a quantitative assessment of the number of messenger RNA copies per cell available for *in situ* hybridization. The data for a coverslip containing 2×10^5 differentiating muscle cells yields a saturation hybridization value of 0.2 ng of probe hybridized at a saturating probe concentration of 2 μg/ml (20 ng/ 10 μl per sample). Furthermore, the saturation value obtained for a fibroblast culture of the same number of cells was 0.1 ng hybridized. These

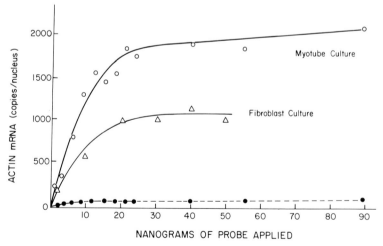

Figure 4-3 The effect of increasing concentration of probe on saturation of target sequences. Hybridization with actin probe was for 3 hours, and the results presented are the average of three samples. The circles represent hybridization to muscle cultures. The open circles represent the actin probe, and closed circles the pBR322 control. The triangles represent hybridization of actin probe to fibroblast cultures. Hybridization of the control pBR322 to fibroblast cultures was essentially the same as for myotube cultures.

results indicate that the optimal concentration of probe is 2 μg/ml, that the use of labeled excess DNA will saturate target sites, and that this saturation level directly depends on the copy number of actin messenger per cell and not some artifactual parameter, such as DNA reannealing.

For the differentiating culture (in which about 20% of the cells are expressing actin in 10-fold higher abundance) the average number of hybridizable actin messages per cell is 2200. For the fibroblast cultures, the average number of messages per cell is 1000. The amount of actin mRNA quantitated by solution hybridization (Schwartz and Rothblum, 1981) was found to be 1800 copies per fibroblast. Hence, we detect more than half of the mRNA molecules present. This shows that the probe is very effective in finding cellular messenger RNA targets (more effective than if the nucleic acids were bound to nitrocellulose; see Stark and Williams, 1979), and therefore *in situ* hybridization can be used as a means of messenger RNA quantitation. The approach yields an absolute estimate of cellular copy number of a particular messenger RNA, assuming a standard efficiency of hybridization. For instance, in fibroblast cultures a vimentin probe hybridized to 40% the level of the actin probe; therefore, after correcting for efficiency of hybridization, there must be about 800 copies of vimentin messenger RNA per cell on the average.

The kinetics of DNA–RNA annealing can be measured by this approach as well. Knowledge of the time course is important to maximize sensitivity of detection. At varying times after introduction of probe to the cells, coverslips can be removed and detected by scintillation counting (Figure 4-4). These results show that hybridization to cellular actin mRNA is detectable in 10 minutes and complete in 4 hours. These results agree with those of Cox et al. (1984) but conflict with many protocols in the *in situ* literature

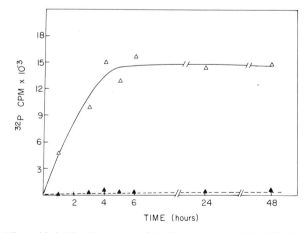

Figure 4-4 Effect of hybridization time on hybridization signal. Hybridization occurred for the times indicated and was performed as described in the text. The open triangles denote actin probe and closed triangles pBR322 control probe. Each point represents the average of six samples.

that call for hybridization times of from 1 to 4 days (Edwards and Wood, 1983; Gee and Roberts, 1983; Haase et al., 1982; Hayashi et al., 1978; Jeffrey, 1982; Singer and Ward, 1982; Venezky et al., 1981; Zimmerman et al., 1983). Hybridization signal sometimes even decreases with time because the hybridization solution contains formamide and high salt, both very good protein solvents, which appear to release cellular RNA. Glutaraldehyde prevents solubilization of RNA over long periods in contrast to formaldehyde. Nevertheless, regardless of the fixative and protocol used, mRNA that is accessible to hybridization by the probe is also likely to be accessible to degradation by any low-level nucleolytic activity. Therefore, we concluded that a hybridization time of approximately 4 hours is optimal, since hybridization in excess of 4 hours would increase the background without increasing in signal. The results obtained here are more similar to the behavior of nucleic acids in solution, and indicate that there is no detectable lag time for the probes to penetrate the cell (we can detect hybrids in less than 10 minutes). Most likely, therefore, the lower efficiency of *in situ* hybridization compared with that of solution hybridization can be explained by messenger ribonucleoproteins obstructing probe accessibility to the target and not by penetration of the probe into the cell.

We have begun to apply a similar quantitative approach for optimization of hybridization to chromosomal and nuclear DNA, the efficiency of which is generally low by existing techniques (see Kozak et al., 1977; Harper et al., 1981; Gerhard et al., 1981). To generate sufficient signal for rapid quantitation of results, we used total human DNA labeled with ^{32}P as a probe for hybridization to standard preparations of chromosomes. Several technical parameters have been tested, two of which will be presented here to illustrate the type of information obtained by this approach.

Standard techniques for *in situ* hybridization to metaphase chromosomes call for long hybridization times (1–4 days) and low concentrations of probe (less than 1 ng/sample). One of the first parameters we tested was the kinetics of the *in situ* hybridization process, and the results of one such experiment are presented in Figure 4-5. It should be noted that each point represents the average of triplicate samples, and the conclusions are generally based on two or three replicate experiments. The striking feature of these results is that hybridization rises very sharply, with the reaction being one-third maximal in just 10 minutes and essentially complete within 4 hours. In some experiments we actually see a slight decrease in signal with prolonged incubations, probably because of dissolution of the sample. Although hybridization of low copy sequences should continue to occur over a longer period of time, we see no evidence of this in time course studies in which hybridization has been monitored for up to 5 days. These results led to the realization that the DNA in the chromosomes, which is melted just before hybridization, will reanneal very rapidly because homologous strands lie so closely together. Hence, the reannealment of chromosomal DNA is most likely the major limiting factor in obtaining good hybridization efficiency to chromosomal or nuclear RNA. It appears that this factor may not have been

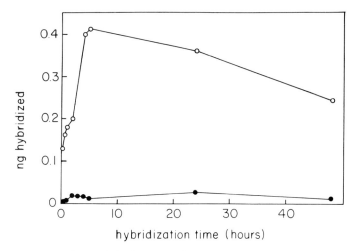

Figure 4-5 Kinetics of *in situ* hybridization to chromosomes. Samples of human chromosome preparations were hybridized at 37° C for the indicated times with 4 ng per sample of ³²P-labeled human genomic DNA (*open circles*). The results were quantitated in a scintillation counter, and each point represents the average of triplicate samples. The closed circles represent background as determined using ³²P-labeled *E. coli* DNA.

considered in standard *in situ* hybridization techniques that call for prolonged hybridization times and low probe concentrations. The logical ramification, supported by the time course results, is that probe concentrations should be as high as possible, so that hybridization of the probe to chromosomal DNA can compete more effectively with the rapid interstrand reannealing of chromosomal DNA.

These conclusions led us to evaluate the effect of probe concentration on signal and noise, since high probe concentrations are generally considered prohibitive because of increased background. Two experiments yielded similar results, and the results of one such experiment is presented in Figure 4-6. Note that hybridization is still rising rapidly at 200 ng/coverslip (10 μg/ml, or over 1000× more DNA than is routinely used in hybridizations), while background is increasing only very gradually. Hence, at high probe concentrations, the extent of hybridization is much greater, while the signal-to-noise ratio can be as good as or better than that at low concentrations. Although the comparable concentrations of cloned probes will be different, the point is that in these experiments hybridization rises more sharply with increasing probe concentrations than does background. These results suggest that better hybridization efficiencies can be obtained by changing *in situ* hybridization strategies to relatively short hybridization times and high probe concentrations.

The rapid reannealing of chromosomal sequences could be eliminated entirely if the nuclear and chromosomal DNA could be fixed immediately after denaturing so that homologous strands remained single stranded. We are currently evaluating a variety of postdenaturation fixation steps to derive a

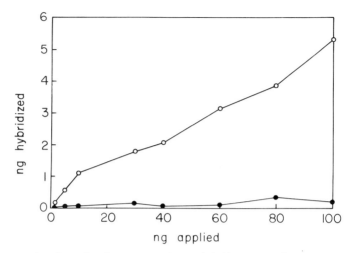

Figure 4-6 The effect of probe concentration on hybridization to chromosomes and background. Samples of human chromosome preparations were hybridized overnight at 37° C and each point represents the average of duplicate samples. Human genomic DNA probe is represented by open circles. Background was determined using *E. coli* DNA as a negative control probe (*closed circles*).

protocol that would sustain the chromosomal DNA in a state in which it is continually available to hybridize with the probe.

Detection of Hybridization

The purpose of the quantitative analytical approach just described was to maximize conditions for hybridization. The full potential of the *in situ* hybridization approach can then be achieved using one of several methods for detecting the hybridized probe directly within individual cells or on chromosomes.

The most common method of detection is the use of autoradiography with tritiated probes. We also have used sulfonated probes with increased convenience (less development time) over tritiated probes, but some loss of resolution is evident. We found that dithiothreitol must be used in the hybridization solution with sulfonated probes to reduce background. In general, we find that radioactive probes currently provide the most sensitive means of detection because exposure time can be extended to enhance signal. One of the advantages of using radioactive probes in conjunction with a photographic emulsion is the ability to obtain quantitative information on single cells by counting the silver grains produced as a result of the radioactive decay. With a 2-month exposure using a tritiated probe with specific activity of 2×10^7 cpm/μg, we can detect 1000 copies of actin message per cell with a signal-to-noise ratio (grains per cell using the actin probe compared with grains per cell using the control probe) of 59:1 comparable with the 70:1 ratio obtained in the ^{32}P experiments. From these results, it was determined that the limit of detection is approximately 30 molecules of the average (2-kb) message per cell, or about 60 kb. This limit is approximated by dividing the signal-to-noise ratio into the number of copies per cell and is based on the assumption that we can detect a signal-to-noise ratio of 2:1 with reasonable accuracy.

A disadvantage of radioactive probes with good resolution, such as tritium, is that a great deal of development time is needed, since specific activity is low and the capture ratio of the weak decay is low in efficiency. The use of sulfonated probes allows much faster development (approximately one tenth the time of tritium), because the energy of decay is much greater and the ^{35}S probe has a higher specific activity. However, a corresponding loss of resolution occurs because the track of the particle is longer.

We have further investigated the use of nonisotopic probes for *in situ* hybridization since our initial work (Singer and Ward, 1982). An evaluation of the biotin analog's effect on all the parameters investigated with isotopically labeled probes showed that the incorporated biotin had no effect on *in situ* hybridization except for its effect on probe size, which was previously discussed. Because biotinylated DNA probes give a consistently higher background, the probe fragment size after nick translation should be 450 nucleotides or less. This means that networking of biotinylated probes will be more difficult than with radioactive probes. This is confirmed by the lack of signal amplification shown in Figure 4-2b. We have found that pretreating the cells with acetic anhydride just before hybridization reduces the nonspecific adherence of biotinylated probes. Nonetheless, the higher background means that the signal-to-noise ratio drops from about 70:1 in our isotopically labeled experiments to about 20:1 with biotin-labeled probes.

We then applied the analytical approach to the detection of nonisotopic probes as well. The biotinylated probe can be detected by indirect immunofluorescence to biotin using rabbit antibiotin (Enzo Biochemical) followed by rhodamine-conjugated antirabbit IgG. Initially, high backgrounds obtained by this method have been reduced primarily through affinity purification of both antibodies and treatment of cells in 0.5% Triton in phosphate-buffered saline (PBS) for 10 minutes prior to antibody staining. An alternative, somewhat more convenient detection method is based on the very high affinity of avidin for biotin. A major advantage to the use of avidin or streptavidin derives from the conjugate-forming capabilities of this molecule, since one avidin can bind four biotins to form a bridge with a biotinylated enzyme (see Figure 4-7). High nonspecific sticking of avidin to cells, however, was a detriment to this detection method. We therefore used ^{125}I-labeled avidin to assess means of reducing the background adherence of this protein to cellular components. Table 4-4 shows the results of experiments in which a variety of parameters were investigated by exposing cells to labeled avidin. The main conclusion of this work was that the presence of PBS promoted high nonspecific sticking of avidin to cells. This was, to some extent, surprising since avidin diluted in PBS was routinely used with suc-

PRINCIPLE OF NON-ISOTOPIC, CYTOCHEMICAL
DETECTION OF *in situ* HYBRIDIZED PROBE

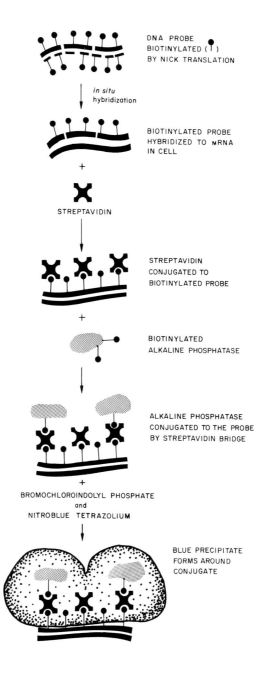

DNA PROBE
BIOTINYLATED (●)
BY NICK TRANSLATION

in situ
hybridization

BIOTINYLATED PROBE
HYBRIDIZED TO mRNA
IN CELL

+

STREPTAVIDIN

STREPTAVIDIN
CONJUGATED TO
BIOTINYLATED PROBE

+

BIOTINYLATED
ALKALINE PHOSPHATASE

ALKALINE PHOSPHATASE
CONJUGATED TO THE PROBE
BY STREPTAVIDIN BRIDGE

+

BROMOCHLOROINDOLYL PHOSPHATE
and
NITROBLUE TETRAZOLIUM

BLUE PRECIPITATE
FORMS AROUND
CONJUGATE

Table 4-4 ^{125}I-Avidin bound to cells with various treatments (average of duplicate samples)

Treatment	Counts per minute
No cells (glass coverslip) 1 × SSC	139
With cells 1 × SSC	1,348
Cells treated with acetic anhydride 1 × SSC	1,365
Triton-treated cells 1 × SSC	1,905
Cells in PBS	16,406
Cells in PBS with 100 μg/ml cold avidin	1,288
Cells in PBS with 1 mg/ml cold avidin	928
Cells in 4 × SSC, pH 5.2	1,174
Cells in 4 × SSC, pH 6.2	844
Cells in 4 × SSC, pH 7.2	1,235
Cells in 4 × SSC, pH 8.2	747

cess in detecting biotinylated probe on filters. The nonspecific sticking of avidin to cells could be reduced by approximately 90% with the use of sodium citrate, preferably of higher salt concentration (4 × SSC). Neither acetic anhydride nor pretreatment with Triton reduced the avidin background. The pH of the solution showed only a small and variable effect on nonspecific sticking of avidin. With hybridization using the maximum probe concentration and saturating the targets so that 100 pg of probe are hybridized per coverslip, and then using avidin with 4 × SSC, a signal-to-noise ratio approximating 15:1 is obtained (data not shown). This indicates that the maximum sensitivity obtainable using the biotin-avidin system would be about 200 molecules per cells of a 2-kb message (the noise of the biotin probe and the avidin system are additive). However, when fluorescence is used as the detector, the signal-to-noise ratio may be reduced further because of autofluorescence of the cells. Autofluorescence is not a significant consideration in hybridization to chromosomes or isolated nuclei and, in our hands, there is extremely little background fluorescence using avidin-fluorescein for detection on such preparations. Recent results from our laboratory indicate that we can consistently detect 20 kb or less in the genome using this method.

The enzymatic detection systems used in conjunction with biotin may hold the most promise for sensitivity (Brigati et al., 1983; Leary et al., 1983; Singer et al., 1986), although their resolution is inferior to that of fluorescence. Either biotinylated peroxidase (Vector Laboratories, Burlingame, CA) or biotinylated alkaline phosphatase (Bethesda Research Laboratories, Gaithersburg, MD) as a complex with avidin serves as a valuable histo-

Figure 4-7 Illustration of histochemical detection of biotinated probes using alkaline phosphatase. In this illustration, the avidin, capable of binding four biotins, is shown forming a bridge between the biotinated probe and the biotinated alkaline phosphatase. In the cells, this must be a two-step procedure in which the avidin is added first, then washed away, and the alkaline phosphatase is added second. This ensures that the conjugate forms in the cells and not outside, where it could not penetrate to the probe.

chemical approach to messenger RNA detection. Because the enzymes can continue to precipitate substrate for long times, the signal is amplified until the precipitate limits substrate diffusion. In our current experiments, we found that the signal-to-noise ratio is substantially improved using this method. However, an exact analysis is not yet feasible because no convenient means of quantitating the enzymatic reaction exists. Figure 4-7 illustrates the principle of the histochemical detection of biotinylated probe. Plate 3a shows the results of this detection when applied to muscle cultures. A probe specific to differentiating muscle myotubes was used in Plate 3a, in which the expression of this gene is evident. The probe contained an untranslated region (3') specific for the cardiac actin isoform (Paterson and Eldridge, 1984), expressed only in the myotubes. This micrograph shows that the myotubes in the field are evident with the histochemical stain. The developing myofibers can be seen with the histochemical staining, and the single undifferentiated cells are only weakly visible. This procedure can be done from beginning to end in a single day, and illustrates the potential of this approach. Plate 3b shows that this procedure can be applied to tissue sections as well. For this illustration we used a 1.8-kb probe to exon 3 of the gene for proopiomelanocortin mRNA kindly supplied by Julie Jonassen (originally cloned by Jim Eberwine) on tissue sections of rat pituitary, also supplied by Julie Jonassen. The dark band of tissue is the intermediate lobe of the pituitary, known to contain high concentrations of POMC RNA (Roberts et al., 1982).

Conclusions

Our goal in this as well as a previous work was to analyze the variables affecting hybridization of nick-translated DNA probes to nucleic acids *in situ*. These studies have allowed us to begin a long process toward assessing various aspects of *in situ* hybridization, which maximize the hybridization signal and the retention of cellular RNA and minimize background. In addition, we have evaluated and compared different methods of detecting DNA hybridized within the cell. The use of this approach allowed us to design an appropriate protocol for *in situ* hybridization to cells in culture that we have applied to tissue sections as well. It should be emphasized that this protocol is not intended to make all preexisting methodologies for *in situ* hybridization obsolete. Because it results from a quantitative evaluation of the factors involved in the procedure, it represents a more fully characterized and streamlined methodology.

Recent work from Angerer's laboratory (Cox et al., 1984) using single-stranded RNA probes for *in situ* hybridization indicated that these probes were more effective in generating hybridization signal than were symmetric (DNA or RNA) probes. We have optimized our hybridization protocol for the use of DNA probes and believe we have avoided procedures that tend to inhibit signal generation from DNA probes (e.g., the use of glutaraldehyde). Although we have not rigorously evaluated this approach using RNA probes, in our hands we found these probes technically more complex and background levels less reproducible. In addition, we favor DNA probes because of their stability, the convenience of not requiring further cloning and subsequent characterization in another vector, the predictable nature of the nick-translation reaction in the generation of fragments of varying sizes, and the low level of nonspecific sticking of the probe. Furthermore, as described previously, our conditions allow the use of the double-stranded probes to form networks, thereby amplifying the signal more than an order of magnitude beyond that which is theoretically obtainable with single-stranded probes. If applied appropriately, the networking of large double-stranded probes may be valuable for detecting low-copy mRNA and viral sequences, and it is likely to be essential for localizing unique genes on chromosomes.

Our results show that fixation is the determining step for subsequent procedures. Paraformaldehyde was found to be the most convenient fixative because it preserves cellular RNA and morphology and allows probes of a wide range of sizes to penetrate the cell without the need of proteolytic digestion. An additional advantage stems from the growing applicability of biotinylated probes, which have even more difficulty penetrating the cellular matrix than other probes. Furthermore, cells exhibit lower autofluorescence, so fluorescent detection systems can be used albeit with less sensitivity at this point. Glutaraldehyde and Zenker's can be used as fixatives with good results as well, provided appropriately small probe fragments are used in conjunction with proteolytic digestion. In general, results from the fixation experiments led us to conclude that development of a successful *in situ* hybridization protocol will require sacrificing certain advantages (e.g., optimal morphology) for the sake of others (e.g., maximal hybridization). Once the cells are fixed properly, the protocol can be used to generate quantitative information on the amount of a nucleic acid within a cell or tissue, as long as the probe size is small in relation to the vector. This method thus allows a rapid, sensitive, and convenient means of quantitation in that many samples can be investigated on a small amount of biological material within a single day. *In situ* hybridization can therefore provide a convenient alternative to nitrocellulose blots or solution hybridization for quantitating average numbers of a nucleic acid per cell.

Although autoradiography remains the most sensitive means of detection, and improvements in speed have been made using sulfonated nucleotides, advances in nonisotopic detection using biotinylated nucleotide analogs are being made that may eventually make these methods preferable to isotopic means. Currently, detection of sequences on nitrocellulose by biotinylated probes are about a picogram of DNA, allowing single-copy detection on Southern blots (in our hands, using the BRL DNA detection kit). Our current work is now directed toward further improving these nonisotopic detection parameters.

When applying our approach to chromosomal detection of *in situ* hybrids, we have used a total genomic probe, to get large signals with which to analyze the various parameters of *in situ* hybridization. Our initial experiments indicate that use of short hybridization times and relatively high probe concentrations may be the most effective means of competing with the rapid reannealment of chromosomal DNA. Work is currently in progress to evaluate this approach using cloned probes for specific sequences.

The work described here illustrates the direction of current *in situ* hybridization methodology. Hopefully, advances being made by a number of laboratories will improve this technology still further.

METHODS

Cell Culture

Skeletal myoblasts were isolated from the pectoral muscle of 12-day-old chicken embryos and cultured by standard techniques. Cells were plated at a density of 2×10^6 per 100-mm plate in minimum essential medium, 10% heat-inactivated fetal calf serum, and 2% chicken serum. Cells were plated into dishes containing glass or plastic coverslips previously autoclaved in 0.5% gelatin. The cells were fixed after 2 to 3 days of incubation, when cultures consisted of a mixture of fibroblasts, undifferentiated myoblasts, and early myofibers. Coverslips containing cells were rinsed twice in Hanks balanced salt solution and fixed for 15 minutes in 4% paraformaldehyde (Fisher) in phosphate-buffered saline with 5 mM MgCl$_2$. To make up the fixative, paraformaldehyde was dissolved in PBS with low heat for 2 to 4 hours, MgCl$_2$ was added and the solution filtered. The fixative can be used for 2 weeks when refrigerated. After fixation, the cells were placed in 70% EtOH at 4° C until later use. A large number of coverslips with cells of uniform density were prepared from each cell culturing, so that parallel samples could be examined with different probes.

Tissue Sections

As described previously (in the section on Fixation), tissues were frozen in isopentane at liquid nitrogen temperatures. The frozen tissue was then cryosectioned at 5 to 8 μm (the thicker sections preserved the morphology better) and then allowed to adhere to histostik-coated coverslips at room temperature for 5 minutes. The coverslips at this point were fixed in 4% paraformaldehyde for 15 minutes, dehydrated through ethanol, and air dried until hybridization. They were then treated the same as for cultured cells. Because the tissue has less surface area than the cells, the washing of the probe and other reagents should be three times longer. Use of biotinylated probes for detection significantly decreases sensitivity because reagents tend to bind more to the tissue. The signal-to-noise ratio measured by scintillation counting (^{32}P) for isotopically labeled actin probe was 15:1, with an average tissue area of 40 μm^2 sectioned at 5 μm. At the present time we are continuing to optimize the technology of nonisotopic detection with tissue sections.

Hybridization to Chromosomes

Preparations of metaphase chromosomes and interphase nuclei were prepared using standard cytogenetic techniques (Kozak et al., 1977) from human HeLa cells. Rapidly dividing monolayer cultures were treated with 0.015 μg/ml of colcemid for 2 to 3 hours and rounded cells were harvested by shaking. Cells were swollen in 0.075 M KCl for 17 minutes at 37° C, fixed in several changes of 3 methanol:1 acetic acid, and dropped at equal densities on glass coverslips. Preparations were hardened overnight at room temperature and then stored at −80° C with dessicant until further use. The protocol for hybridization to chromosomes represents a modification of previously described techniques (Harper et al., 1981, Gerhard et al., 1981). Before hybridization, samples were digested with 100 μg/ml RNase A (Sigma) in 2 × SSC; samples were then incubated for 10 minutes in 0.1 M triethanolamine containing 0.25% acetic anhydride for reduction of background. Chromosomal DNA was denatured by incubation at 70° C for 2 minutes in 70% formamide, at 2 × SSC. Samples were then dehydrated through cold 70, 95, and 100% EtOH and air dried. The hybridization buffer was the same as that described later except that vanadyl sulfate was omitted and probe concentrations and incubation times were varied as indicated. The samples were rinsed in 50% formamide, at 2 × SSC for 30 minutes and then 1 × SSC for 30 minutes. Hybridization of ^{32}P-labeled probes to samples in PBS was quantitated by Cerenkov radiation in a scintillation counter.

Probes and Nick Translation

The actin probe consisted of a full-length (2-kb) transcript coding region for chicken beta-actin inserted into pBR322 (Cleveland et al., 1980). This probe hybridizes with the mRNAs of different actin isoforms under the hybridization conditions employed. The control probe used was pBR322 without any insert. Plasmid DNA was routinely nick-translated using ^{32}P-dCTP (Amersham) or three ^{3}H-labeled nucleotide triphosphates (New England Nuclear) 54 to 100 Ci/mmol. Specific activity of ^{3}H-labeled probes ranged from 1 × 10^7 to 3 × 10^7 cpm/μg. For probes nick-translated with ^{35}S dCTP, the specific activity ranged from 1.7 × 10^8 to 2.9 × 10^8 cpm/μg. For nonisotopic detection, probes were labeled by nick translation with biotinylated dUTP (Enzo Biochemical). Probe fragment size was controlled by varying the amount of DNase (Worthington) in the nick-translation reaction, and the size was monitored using 1.5% alkaline agarose gel electrophoresis. We found the most useful DNase concentration to be approximately 10 to 20 pg/ml final. Biotinylated probes were sized using alkaline phosphatase detection of probes transferred to nitrocellulose (see Detection of Hybridization).

Hybridization

Cells fixed in paraformaldehyde (or tissue dried through ethanol) and stored in 70% ethanol, as indicated previously, were rehydrated in phosphate-buffered saline plus 5 mM MgCl$_2$ for 10 minutes, and then in 0.1 M glycine, 0.2 M Tris HCl, at pH 7.4 for 10 minutes. The samples were then placed in 50% formamide, 2 × SSC (0.3 M sodium chloride in sodium citrate buffer) for 10 minutes at 60° C before hybridization. The probe, tRNA, and salmon sperm DNA were lyophilized and then resuspended in formamide and melted at 90° C for 10 minutes. Just before it was placed on the samples, the probe was combined with a hybridization mix so that the final probe concentration was 1 to 2 µg/ml and the final hybridization solution consisted of 50% formamide (Fluka, deionized with mixed bed resin, BioRad AG501-XB for 1/2 hour, then filtered), 2 × SSC, 0.2% BSA, 10 mM vanadyl sulfate ribonucleoside complex (Berger and Birkenmeier, 1979), 10% dextran sulfate (Sigma), and 1 mg/ml final concentration each of *E. coli* tRNA and sheared salmon sperm DNA. For hybridizations with ^{35}S, DTT was added to the hybridization solution (10 mM final, R. Angerer, personal communication). Samples on coverslips were incubated in hybridization solution for 4 hours at 37° C by putting the coverslips cell side down onto 20 µl of the solution on Parafilm. After hybridization, the coverslips were placed in small Coplin jars (VWR) and rinsed three times for 30 minutes each in 2 × SSC, 50% formamide at 37° C; 1 × SSC, 50% formamide at 37° C, and 1 × SSC at room temperature while being shaken.

Detection of Hybridization

Coverslips hybridized with ^{32}P probes were cut in half with a diamond pencil and counted by Cerenkov radiation in a scintillation counter. Samples hybridized with [3]H- or ^{35}S-labeled probes were dehydrated through 70, 95, and 100% ethanol and air dried. Coverslips were mounted cell side up on slides using Permount and dipped into Kodak NTB-2 emulsion. Air-dried slides were placed in light-tight boxes with Drierite and stored at 4° C. Exposure times for ^{3}H-labeled probes were from 8 to 12 weeks, and those for ^{35}S-labeled probes were from 3 to 12 days. Slides were processed through D-19 developer for 5 minutes, 1% acetic acid for 30 seconds, Kodak fixer for 5 minutes, and a water rinse for 30 minutes. Slides were immersed in a freshly prepared solution of 5% Giemsa stain (BDH Chem.) for 20 to 30 minutes in 0.05 M Tris HCl, pH 6.8. Alternatively, the slides were stained with the DNA fluorochrome DAPI (4'–6'diamidino-2-phenylindole) for 10 minutes at 1 μg/ml in PBS.

For detection of probes labeled with biotin, two alternative techniques were employed. For fluorescence detection, samples hybridized with biotinylated probe were reacted with 2 μg/ml of rhodamine-avidin (Vector Laboratories) in 1% BSA, 4 × SSC for 10 minutes (Singer and Ward, 1982). Samples were then rinsed for 1 hour in 4 × SSC, mounted in antibleach mounting medium, and viewed with epifluorescence optics. However, good results are obtained using an antibody against biotin made in rabbit (Enzo Biochemicals) and a second antibody, which has been rhodamine conjugated. Blocking nonimmune serum from the second animal should be used at a concentration of 1%. The antibodies are used at 2 μg/ml in 2 × SSC, and incubations last for 20 minutes at 37° C. The cells are then washed in three changes of 2 × SSC, with shaking, at room temperature. The fluorochrome can then be viewed after the coverslips are mounted in Tris-glycerol on microscope slides using a bleaching inhibitor (Johnson and Nogueira, 1981). Alternatively, biotinylated probes were detected by a colorimetric reaction using alkaline phosphatase (Leary et al., 1983). Samples hybridized with biotinylated probes were reacted with streptavidin and then with a biotinylated-alkaline phosphatase complex. The streptavidin provides a bridge between the biotinylated DNA and the biotinylated enzyme (see Figure 4-7). Alkaline phosphatase is then detected by incubation of cells in a mixture of 5-bromo-4-chloro-3-indolyl phosphate and nitroblue tetrazolium, which products a dark purple precipitate at sites of hybridization. The details of this procedure follow.

The coverslips were removed from ethanol, rehydrated in Tris-glycine, and then incubated at 42° C for 10 minutes in a solution containing 0.1 M Tris HCl, pH 7.5, 0.1 M NaCl, 2 mM MgCl$_2$, 0.05% Triton X-100, and 3% BSA. They were then placed onto 20 μl of streptavidin (0.5–1 μg/ml) in 4 × SSC, 0.5% Triton, and 2 mM MgCl$_2$ for 10 minutes on Parafilm. They were washed once in 4 × SSC and twice in 0.1 M Tris HCl, pH 7.5, 0.1 M NaCl, 2 mM MgCl$_2$, and 0.05% Triton X-100 for 5 minutes each. They were then incubated with biotinylated alkaline phosphatase (1 μg/ml) in the same solution for 10 minutes at room temperature, in the same manner as the streptavidin. The samples were washed with three changes of the solution, and a final rinse in 0.1 M Tris HCl, at pH 9.5, 0.1 M NaCl, and 50 mM MgCl$_2$ for 2 minutes. The samples were then put in a solution of 330 μg/ml of nitroblue tetrazolium, mixed with bromochloroindolyl phosphate just before its use at a final concentration of 166 μg/ml. The nitroblue tetrozolium is kept as a stock at 75 mg/ml in a solution of 70% dimethyl formamide, and the bromochloroindolyl phosphate is kept at 50 mg/ml in a solution of anhydrous dimethyl formamide for no more than 2 weeks in the freezer. All the reagents are available in kit form from Bethesda Research Laboratories (DNA detection kit, Catalog No. 82395A). The biotinylated nucleotides can also be detected on nitrocellulose by similar methods with a sensitivity comparable to a 1-day autoradiographic exposure.

Note added in proof: The authors would like to acknowledge the work of E. R. Unger, L. R. Budgean, O. Myerson, and D. J. Brigati (*Am. J. Surg. Path.* 10:1–8) showing that alkaline phosphatase can be used to detect biotinylated probes.

ACKNOWLEDGMENTS

We express appreciation to David Ward for his valuable advice and contributions to the alkaline phosphatase detection. We acknowledge Carol Villnave for her excellent technical assistance. Julie Jonassen very generously allowed us the use of the POMC probe and her tissue sections of the pituitary. We appreciate the fine secretarial work of Elayn Byron and photographic work of Marie Giorgio. This work was supported by NIH grant HD18066 and grants to RHS and JBL from the Muscular Dystrophy Association.

REFERENCES

Angerer, L.M. and Angerer, R.C. Detection of poly A^+ RNA in sea urchin eggs and embryos by quantitative *in situ* hybridization. *Nucleic Acids Res.* 9:2819–2840, 1981.

Berger, S.L. and Birkenmeier, C.S. Inhibition of intractable nucleases with ribonucleoside-vanadyl complexes: Isolation of messenger ribonucleic acid from resting lymphocytes. *Biochemistry* 18:5143–5149, 1979.

Brigati, D.J., Myerson, D., Leary, J.J., Spalholz, B., Travis, S.Z., Fong, C.K.Y., Hsiung, G.D., and Ward, D.S. Detection of viral genomes in cultured cells and paraffin-embedded tissue sections using biotin-labeled hybridization probes. *Virology* 126:32–50, 1983.

Capco, D.G. and Jeffrey, W.R. Differential distribution of poly(a)-containing RNA in the embryonic cells of *Oncopeltus fasciatus*. Analysis by *in situ* hybridization with a [3H] poly (U) probe. *Devel. Biol.* 67:137–151, 1978.

Cleveland, D.W., Lopata, M.A., McDonald, R.J., Cowan, N.K., Rutter, W.J., and Kirschner, M.W. Number and evolutionary conservation of alpha- and beta-tubulin and cytoplasmic beta- and gamma-actin genes using specific cloned cDNA probes. *Cell* 20:95–105, 1980.

Cox, K.H., DeLeon, D.V., Angerer, L.M., and Angerer, R.C. Detection of mRNAs in sea urchin embryos by *in situ* hybridization using asymmetric RNA probes. *Devel. Biol.* 101:485–502, 1984.

Edwards, M.K. and Wood, W.B. Location of specific messenger RNAs in *Caenorhabditis elegans* by cytological hybridization. *Dev. Biol.* 97:375–390, 1983.

Gee, C.E. and Roberts, J.L. *In situ* hybridization histochemistry: A technique for the study of gene expression in single cells, In *DNA 2*. Mary Ann Liebert, Inc., Publishers, 1983, pp. 157–163.

Gerhard, D.S., Kawasaki, E.S., Carter Bancroft, F., and Szabo, P. Localization of a unique gene by direct hybridization *in situ. Proc. Natl. Acad. Sci. USA* 78:3755–3759, 1981.

Haase, A.T., Stowring, L., Harris, J.D., Traynor, B., Ventura, P., Peluso, R., and

Brahic, M. Visna DNA synthesis and the tempo of infection *in vitro*. *Virology* 119:399–410, 1982.

Haase, A.T., Walker, D., Stowring, L., Ventura, P., Geballe, A., and Blum, H. Detection of two viral genomes in single cells by double-label hybridization *in situ* and color microradioautography. *Science* 227:189–191, 1985.

Harper, M.E., Ullrich, A., and Saunders, G.R. Localization of the human insulin gene to the distal end of the short arm of chromosome 11. *Proc. Natl. Acad. Sci. USA* 78:4458–4460, 1981.

Hayashi, S., Gillam, I.C., Delaney, A.D., and Tener, G.M. Acetylation of chromosome squashes of *Drosophila melanogaster* decreases the background in autoradiographs from hybridization with [^{125}I]-labeled RNA. *J. Histochem. Cytochem.* 26:677–679, 1978.

Jeffrey, W. Messenger RNA in the cytoskeletal framework: Analysis by *in situ* hybridization. *J. Cell Biol.* 95:1–7, 1982.

Johnson, G.D. and Araujo Nogueira, G.M. A simple method of reducing the fading immunofluorescence during microscopy. *J. Immunol. Methods* 43:349–350, 1981.

Kozak, C.A., Lawrence, J.B., and Ruddle, F.H. A sequential staining technique for the chromosomal analysis of interspecific mouse/hamster and mouse/human somatic/human somatic cell hybrids. *Exp. Cell Res.* 105:109–117, 1977.

Lawrence, J.B. and Singer, R.H. Quantitative analysis of *in situ* hybridization methods for the detection of actin gene expression. *Nucleic Acids Res.* 13:1777–1799, 1985.

Lawrence, J.B. and Singer, R.H. Intracellular localization of messenger RNA for cytoskeletal proteins. *Cell* 45:407–415, 1986.

Leary, J.J., Brigati, D.J., and Ward, D.C. Rapid and sensitive colormetric method for visualizing biotinylated-labeled DNA probes hybridized to DNA or RNA immobilized on nitrocellulose: Bio-blots. *Proc. Natl. Acad. Sci. USA* 80:4045–4049, 1983.

Paterson, B.M. and Eldridge, J.D. Alpha-cardiac actin is the major sarcomeric isoform expressed in embryonic avian skeletal muscle. *Science* 224:1436–1438, 1984.

Roberts, J.L., Chen, C.C., Eberwine, J.H., Evinger, M.J.Q., Gee, C., Herbert, E., and Schachter, B.S. Glucocorticoid regulation of proopiomelanocortin gene expression in rodent pituitary. *Recent Prog. Horm. Res.* 38:227–256, 1982.

Schwartz, R.J. and Rothblum, K.N. Gene switching in myogenesis: Differential expression of the chicken actin multigene family. *Biochem.* 20:4122–4129, 1981.

Singer, R.H. and Kessler-Icekson, G. Stability of polyadenylated RNA in differentiating myogenic cells. *Eur. J. Bioch.* 88:395–407, 1978.

Singer, R.H. and Ward, D.C. Actin gene expression visualized in chicken muscle tissue culture by using *in situ* hybridization with a biotinylated nucleotide analog. *Proc. Natl. Acad. Sci. USA* 79:7331–7335, 1982.

Singer, R.H., Lawrence, J.B., and Villnave, C. Optimization of *in situ* hybridization using isotopic and non-isotopic detection methods. *Biotechniques* 4:230–250, 1986.

Stark, G.R. and Williams, J.G. Quantitative analysis of specific labelled RNA's using DNA covalently linked to diazobenzyloxymethyl-paper. *Nucleic Acids Res.* 6:195–203, 1979.

Venezky, D.L., Angerer, L.M., and Angerer, R.C. Accumulation of histone repeat transcripts in the sea urchin egg pronucleus. *Cell* 24:385–391, 1981.

Vincent, A., Civelli, O., Maundrell, K., and Scherrer, K. Identification and characterization of the translationally repressed cytoplasmic globin messenger-ribonucleoprotein particles from duck erythroblasts. *Eur. J. Biochem.* 112:617–633, 1980.

Zimmerman, J.L., Petri, W., and Meselson, M. Accumulation of a specific subset of *D. melanogaster* heat shock mRNAs in normal development without heat shock. *Cell* 32:1161–1170, 1983.

5

Methods for Hybridization and Quantitation of mRNA in Individual Brain Cells

W. Sue T. Griffin

In situ hybridization techniques developed to quantitate the levels of specific messenger RNAs (mRNAs) relative to the levels of total polyadenylated (poly-A) mRNA are detailed here together with our protocols for combining immunohistochemical localization of proteins encoded by mRNAs that are localized by *in situ* hybridization. Specifically, I will demonstrate the use of the protocols we have used for *in situ* hybridization of cDNA, synthetic oligodeoxynucleotides, and RNA probes (Griffin et al., 1983; Griffin et al., 1985; Griffin and Morrison, 1985) radiolabeled with either [^{3}H], [^{35}S], or [^{32}P], to paraffin-embedded mammalian tissue preserved with a variety of fixatives. Because many of those whose work has led to the use of *in situ* hybridization as a laboratory technique are the authors of other chapters in this volume (Jones, 1973; Harrison et al., 1974; Brahic and Haase, 1978; Capco and Jeffrey, 1978; Angerer and Angerer, 1981; Singer and Ward, 1982), the number of references cited will be kept to a minimum.

METHODS

Tissue Preservation

The primary motive of tissue preservation for *in situ* hybridization is to stabilize mRNA within the cellular matrix in a manner that leaves it accessible for hybridization. For quantitation of the relative levels of a specific mRNA within a cell type, preservation of structural integrity is also essential. We have used a number of fixatives for hybridization studies, including Bouin's

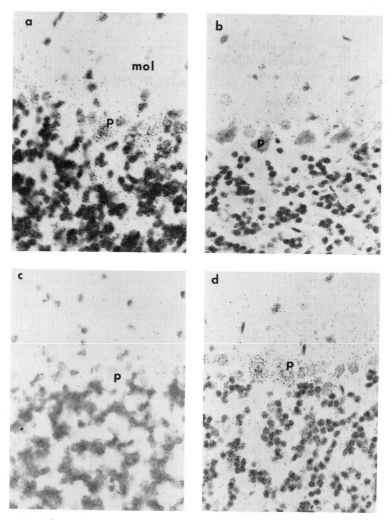

Figure 5-1 These photomicrographs demonstrate hybridization of poly-A RNA in adult rat cerebellum preserved with different fixatives. [^{3}H]-poly-U (100K cpm/section) was hybridized (1 hour at 50° C) by the protocol in Table 5-1. All the slides were processed at the same time and exposed to emulsion for 5 days. Duplicate slides that were pretreated with RNase A in 0.5 × SSC and hybridized at the same time had virtually no autoradiographic grains: (a) was fixed 48 hours in Bouin's solution; (b) was perfused with 4% paraformaldehyde; (c) was placed in SSC and microwaved for 60 seconds; and (d) was placed in Steffanini's (1967) fixative for 3 hours at 4° C.

(Figure 5-1a), neutral buffered formalin, perfused 4% paraformaldehyde (Figure 5-1b), microwave (Figure 5-1c), Steffanini's (Figure 5-1d), perfused periodate-lysine-paraformaldehyde (Figure 5-3), and perfused 2% paraformaldehyde—1% glutaraldehyde (Figure 5-4). Each of these effectively preserves poly-A mRNAs, but some (e.g., microwave) preserve structural integrity less effectively. Directions for preparing these fixatives can be found in standard histology texts; one recently published by Vacca (1985) is particularly useful. The protocol for *in situ* hybridization is outlined in Table 5-1.

Depending on the experimental paradigm, it may be preferable to perfuse the brain (as was the case in Figures 5-1b, 5-3, and 5-4), whereas in others—in which many samples must be taken in a relatively short time or tissue availability is limited, such as human brain samples—perfusion may

Table 5-1 Poly-U *in situ* protocol

1. *Rehydrating sections*
   Place carrier with slides in toluene 5 minutes, 3×; 20 dips: 100% EtOH, 2×; 95%, 2×; 70%.

2. *Permeabilize*
   HCl (0.2 N, 10 minutes); 0.01% Triton X-100 (15 minutes); proteinase K (1 μg/ml, 10 minutes); vary times if necessary.

3. *Pretreatment for electrostatic charge*
   Triethanolamine (10.66 ml/800 ml DW, pH 8 with concentrated HCl). Immerse slides and add 4 ml of acetic anhydride while vigorously stirring. When droplets are dispersed, stop stirring and leave the slides 10 minutes. Follow with 5 minutes 2 × at SSC.

4. *Hybridization*
   Remove slides one at a time and dry around the sections. Cover the sections with hybridization mixture (should form a droplet that remains throughout hybridization without drying out) and place slides on wet paper towels in a box with a lid. When all the slides are ready, cover the chamber and incubate for 1 hour at 50° C.

   Hybridization mix:

Compound	μl
20 × SSC	10
Poly-U 0.05 mCi/ml	10
Sterile DW	80

5. *Posthybridization treatment*
   Cover the sections with the same amount as in the hybridization mixture of RNase A (50 μg/ml of 4 × SSC); incubate in humid chamber for 1 hour at room temperature at 2 × SSC with 0.01% Triton X-100 for 30 minutes on a shaker, followed by two changes of 2 × SSC on the shaker 30 minutes.

6. *Dehydration*
   20 dips each: 70%, 90 (2×), 100 (2×), air dry.

7. *Autoradiographic dipping*
   In a dark room with a safe light, warm (40° C) the emulsion and dilute 1:1 with sterile DW. Mix with a clean slide (avoid bubbles). Dip the slides and stand them upright in test tube racks. Store in light-tight warm, humid chamber for 2 hours, then in black plastic slide boxes in a light-tight refrigerator for an appropriate times (about 4 days for 50K cpm/ 10 μl).

8. *Developing autoradiographs*
   Prepare Kodak D-19 and fixer according to package directions. Into three staining dishes, pour developer, distilled water, and fixer and cool to 16° C. Develop slides 1 minute in the developer, stop in distilled water for 30 seconds, and fix in fixer 3 minutes. Stain.

be less practical. In the latter case, fixation by placing freshly excised tissue into either Bouin's or neutral buffered formalin is the method of choice. The duration the tissue remains in Bouin's fixative is important, as we show in Figure 5-2a–f. Tissue sections from the cerebellum of a 6-day-old rat that was fixed in Bouin's for 30 minutes has a higher background than that fixed for 2 to 48 hours, but the longer fixation times yield higher grain density over cells relative to extracellular spaces. Fixation for 3 days to 3 weeks resulted in no detectable hybridization, indicating that fixation times affect the probability of hybrid formation.

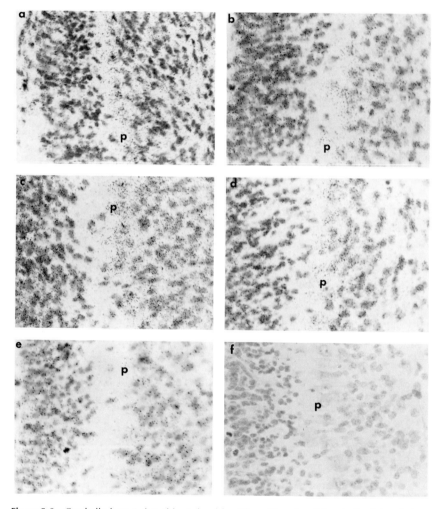

Figure 5-2   Cerebella from 6-day-old rats fixed for different lengths of time in Bouin's solution were hybridized to [³H] poly-U, as in Figure 5-1: (a) was in Bouin's for 30 minutes, (b) for 2 hours, (c) for 4 hours, (d) for 24 hours, (e) for 48 hours, and (f) for 3 days. The abbreviations are as in Figure 5-1.

Preparation of Microscope Slides

Tissue adherence to the slide often poses a significant problem. We find that rigorous cleaning of the slides (detergent washing, with hot water rinsing, followed by alcohol/acetone treatment) before application of a coating solution to facilitate tissue section adherence is beneficial. We have coated slides with Denhardt's (1966) solution (0.02% ficoll, polyvinyl pyrrolidone, and BSA), poly-L-lysine periodate, and gelatin (0.25% gelatin and 0.05% chrom-alum, and 0.3% formalin). The gelatin solution most effectively adheres to the section with high-temperature (above 75° C) low-salt (0.01%) washes. Another criterion of an acceptable coating (also referred to as subbing) solution is that it must not stick to nucleotide strands; neither of the three coating solutions does.

Embedding and Sectioning

For preparation of frozen sections, we have embedded tissue blocks in Tissue Tek (Scientific Products). For better morphology and convenience, paraffin (Figures 5-1, 5-2, 5-4, and Plate 4) and methacrylate (Figure 5-3) are superior. For optimal grain localization, thinner sections (e.g., 2–3 μm obtainable with methacrylate) give better resolution, but for practical reasons, such as availability of human tissue in paraffin blocks, we have concentrated on devising *in situ* protocols for paraffin, which we routinely section at 5 μm. The sections are dried to the slide in either a 37° C oven overnight or at 56 to 58° C for 1 hour and stored in slide cabinets until they are used. Storage at room temperature for over 1 year has not detectably altered hybridization.

Permeabilization

In an attempt to facilitate access of the nucleotide sequence to the mRNA, cells are permeabilized. Permeabilization increases detectable hybridization

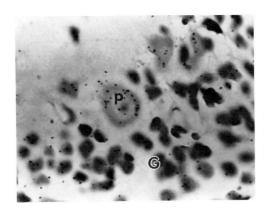

Figure 5-3 Cerebellum from adult rat perfused with periodate-lysine-paraformaldehyde was embedded in methacrylate (Beckstead and Bainton, 1980) and hybridized and exposed as in Figure 5-1. The abbreviations are as in Figure 5-1.

in cells that are not cross-sectioned when the tissue is cut; in addition, hybridization in cells that are cross-sectioned is enhanced. After paraffin sections are deparaffinized (toluene, 3×) and rehydrated (100% ethanol, 2×; 95%, 2×; and 70%), they are permeabilized in either 0.2 N HCl for 10 to 20 minutes, 0.01% Triton X-100 for 1 to 10 minutes or proteinase K (0.5–1.0 μg/ml) for 1 to 10 minutes, or a combination of these. Bouin's fixed neonatal rat brain is treated for 20 minutes with HCl only, whereas adult human brain fixed in either Bouin's or 10% neutral buffered formalin is better permeabilized when all three treatments are applied for 10 minutes each. These permeability regimens are determined empirically; the neonatal Bouin's fixed brain tissue integrity and hybridization was severely compromised by the more aggressive permeabilization schedule, without which hybridization was undetectable in adult, formalin-fixed brain tissue. Bouin's fixed neonatal organs such as spleen and liver may be permeabilized similarly to adult tissue. We have successfully used other permeabilization strategies such as freezing and thawing (Manuelidas and Ward, 1984), which could be advantageous in some applications.

After permeabilization the sections are treated with acetic anhydride solution (5% acetic anhydride in 0.1 M triethanolamine HCl, pH 8.0) for 10 minutes. This treatment reduces electrostatic charge that could cause the probe to bind to the section nonspecifically.

Hybridization

Mixture and Processing

Hybridization mixtures (see Tables 5-1, 5-2, and 5-3) vary with probe-mRNA homology, the length of the probe, cytosine-guanine (CG) content, the rarity of the mRNA, probe concentration, the radionuclide used for labeling, and according to whether the probe is a DNA or RNA sequence. Salt concentrations are higher when heterologous, shorter, lower CG content, or RNA probes are used. To radiolabeled cDNA probe at a final concentration of 10K cpm/μl, we add $E.$ $coli$ DNA and poly-A (500 μg each per milliliter of 2 × SSC:0.15 M NaCl; 0.015 M trisodium citrate). We add formamide up to 50% for homologous probes and up to 25% for heterologous probes. Total yeast RNA, instead of $E.$ $coli$ DNA, is added to cRNA probe mixtures. When the radiolabel is [^{35}S], 0.3 M dithiothreitol is added to the mixture. The 1.5-ml centrifuge tube containing the probe, water, and $E.$ $coli$ DNA or total yeast RNA is placed in a boiling water bath for 2 minutes, after which the other components of the mixture are added and placed on ice.

Of this probe mixture labeled with either [^{3}H] or [^{35}S], 10^5 cpm/10 μl is applied to one of two sections on each slide; the other section is hybridized with an irrelevant probe (e.g., pBR322) of similar length. The slides are then placed in an incubation chamber (a clear plastic, hinged-lid box lined

Table 5-2 *In situ* hybridization

1, 2, 3. *Pretreat slides bearing sections as for poly-U hybridization*

4. *Hybridization*

   Remove the slides one at a time and dry around the sections. Cover the section with the hybridization mixture (should form a droplet that remains throughout hybridization without drying out) and place the slides on wet paper towels in a humid chamber. When all the slides are ready, cover the chamber and incubate. Time and temperature vary. Start with 4 hours at 37° C.

   Hybridization mix:

Compound	μl
20 × SSC	10
Probe	5
Poly-A	10
E. coli DNA	10
Formamide	25
Sterile DW	40
Total volume	100

   n.b. *Stock Solutions* (all kept frozen)

   E. coli DNA 5 mg/ml
   Poly-A 5 mg/ml
   Formamide (BRL)
   Probe is usually 100K cpm/10μl
   Boil probe, DW, and E. Coli 2 minutes; then add
   remaining solutions to the mixtures.

5. *Posthybridization*

   2 × SSC with 0.01% Triton X-100 on shaker 30 minutes (400-ml staining dish); 2 × SSC for 30 minutes on shaker; 0.1 × SSC for 30 minutes on shaker.

---

on the bottom with wet paper towels) and brought up to hybridization temperature in an oven. The slides can be coverslipped to prevent dehydration during incubation by placing an O-ring around the section. However, at a temperature of 50° C or below, dehydration in our chambers sealed with plastic wrap is not a problem.

Temperature

The optimum temperature of hybridization is usually determined empirically, beginning with a temperature 25 degrees Celsius below the T_m of the probe used. The T_m is the temperature at which 50% of the hybrids formed melt apart and can be calculated ($T_m = 0.41 \times \%CG + 69.3° C$) if the CG content is known; if it is unknown, T_m can be estimated by assuming that CG constitutes 50% of the probe. The temperature of hybridization is adjusted for the percentage of formamide added. For each percentage increase in formamide content, the temperature may be decreased by 0.5° C for cDNA probes and 0.35° C for cRNA probes.

Time

The time of hybridization is related to the rarity of the mRNA, the more rare mRNAs require longer hybridization. However, mRNAs classified as rare in solution may be relatively abundant in the cell type synthesizing them. It is therefore necessary to determine the optimal hybridization time for each probe. The radionuclide used also affects the duration of hybridization necessary to detect hybrid formation autoradiographically; for example, the higher-energy beta particle emitted by [^{35}S] is more readily detected than is the lower-energy beta particle emitter by [^{3}H].

Posthybridization Washing

The stringency of washing depends on probe–mRNA homology and can be increased by adding detergents (0.01% Triton X-100), by decreasing the salt concentration, and by increasing the time, temperature, and volume of the wash solution. The apparent T_m is established during posthybridization washing by stepwise increases in the wash temperature, over and above that usually employed, until the number of autoradiographic grains detected per cell is one half the maximum detected. Continuing to increase the temperature above the T_m will melt all the hybrids—further evidence that the autoradiographic grains represent specific DNA–mRNA hybrids. When RNA probes are used, the sections are treated after hybridization with RNase in 4 × SSC to digest single-stranded RNA. Posthybridization treatment with RNase in low-salt concentrations (0.5 × SSC) digests single- and double-stranded RNA—another test of cRNA–mRNA hybridization.

Exposure to Emulsion

After washing, slides are dehydrated through the graded alcohol series and air dried. They may be stored for a few days, if kept dust free. We routinely use Kodak NTB-2 nuclear track emulsion, but other emulsions such as NTB-3 and Ilford K2, K5, and L4 may be used, depending on the particular application; for example, smaller grains develop with NTB-3, K5, and L4. The emulsion should be melted in its container in a water bath (38–40° C) kept in a safe-lit room. Heating the emulsion to higher temperatures can increase the number of background grains and result in uneven coating of the slides. Emulsion may be aliquoted from the original container into reclosable plastic slide carriers (Scientific Products). The containers are filled to one-third capacity, wrapped in black paper, and stored light-tight at 4° C until they are diluted 1:1 with sterile distilled water just before their use. Emulsion may be used undiluted, reheated numerous times, and stored in its original box at 4° C, usually without adverse effect; we aliquot rather than reuse emulsion, however, in an attempt to circumvent problems with emulsion-related background autoradiographic grains. Also, pressure on the flexible sides of the carriers allows more slides to be coated from a given volume of emulsion. The diluted emulsion is stirred with several dips of a clean slide; any resulting bubbles are allowed to dissipate. Dipping the slides bearing sections slowly and smoothly into the carriers results in a thin, even coating of the sections. A bright safe light such as a sodium vapor lamp (Thomas) is helpful in detecting the formation of bubbles on the sections, which can result in uneven thickness of the emulsion, if not chemographic artifacts, and in ensuring that the upper section is coated with the emulsion. After coating, the slides are placed in a rack and stored for 2 hours in a humidified, light-tight cabinet; a waterbath inside the cabinet is heated to 40° C and then turned off just before the slides are stored. If the temperature inside the box is too high (above approximately 40° C), the emulsion will crack; and if the humidity is above approximately 90%, the emulsion will melt off the slide. After the emulsion has set on the slides in the humid cabinet, the slides are placed in plastic slide boxes containing a small vial filled with dessicant and stored in a light-tight box at 4° C.

The length of time the radiolabeled hybrids expose the emulsion is related to the specific activity of the probe (Curie or disintegrations per minute [dpm]/ gram of DNA or RNA), the abundance of the mRNA, and the conditions of hybridization. Our specific activity averages 10^7 cpm/μg using [^{3}H] for nick translation. Hybridizing to a cell-specific, relatively abundant mRNA such as proopiomelanocortin (POMC) in the intermediate lobe of the pituitary, we expose for 2 weeks (Figure 5-4a). Slightly lower specific activities

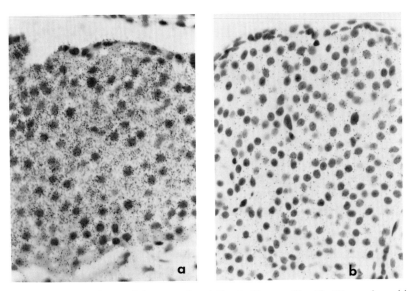

Figure 5-4 Pituitary from adult rat perfused (by B. Chronwall) with 2% paraformaldehyde, 1% glutaraldehyde and embedded in paraffin was hybridized as in Table 5-2 with [^{3}H] POMC cDNA (donated by Jim Roberts) (a) or [^{35}S] POMC synthetic 21 deoxynucleotide sequence (b) and exposed for 5 days.

are achieved by kinasing a synthetic 21-deoxynucleotide sequence (21 mer) of the POMC coding sequence with [^{35}S-gamma-ATP], but because detection is increased with [^{35}S] (a higher-energy beta emitter than [^{3}H]), exposure time can be reduced, even when hybridization times and stringencies are similar to those employed with the [^{3}H]-POMC cDNA probe (Figure 5-4).

Staining

Immunohistochemical Staining

If an antibody is available against the antigen whose expression is being analyzed by *in situ* hybridization, staining by immunohistochemistry permits identification of the cell that synthesizes the protein as well as the cellular localization of the protein. This is particularly important in the nervous system, where proteins may be transported long distances to process terminals or secreted and taken up by other cells. To test the feasibility of combining *in situ* hybridization and peroxidase–antiperoxidase immunohistochemistry (diaminobenzidene was the chromagen; see Table 5-3 for our protocol, which is a modification of the Sternberger, 1979 technique) on Bouin's fixed, paraffin-embedded brain tissue, we *in situ* hybridized [^{3}H]-poly-U to 5-μm sections of adult rat cortex and then localized glial fibrillary acidic protein (GFAP) immunoreactivity in the same section (Plate 4a; Table 5-3). Combining these techniques for colocalization of an mRNA and its translation product in the same cell provides evidence that the cDNA probe is specific to the mRNA and is hybridizing to it in the tissue section. However, the use of adjacent sections may be more practical for quantitative *in situ* hybridization, because the stained immunoreactive product could interfere with the exposure of the emulsion by beta particles, making it imperative to perform both techniques independently to provide baseline data for grain counting.

Table 5-3 Combining *in situ* hybridization and immunohistochemistry after posthybridization washes

---

A. *Blocking endogenous peroxidase*
   1. 100% methanol:3% hydrogen peroxide 1:4, for 30 minutes at room temperature (RT).
   2. Place in TBS for 5 minutes.
   3. Dry slides around sections.

B. *Blocking nonspecific binding of secondary antibody*
   1. Cover section with normal serum from secondary antibody source (e.g., normal goat serum [NGS]) diluted 1:5 with Tris-buffered saline (TBS).
   2. Incubate for 1 hour at room temperature.
   3. Dry slides around sections.

C. *Primary antibody incubation*
   1. Cover section with antibody (appropriately diluted with 2% NGS in TBS).
   2. Incubate for the appropriate time at appropriate temperature.
   3. Run TBS over section, then incubate for 5 minutes in TBS.
   4. Dry slide around sections.

Table 5-3 (continued)

D. *Secondary antibody incubation*
1. Cover section with normal serum from secondary antibody source (e.g., NGS) diluted 1:5 with TBS.
2. Incubate for 30 minutes at room temperature.
3. Dry around sections.

E. *Tertiary antibody incubation*
1. Cover section with primary antibody source peroxidase-antiperoxidase (e.g., rabbit PAP, 1:300 in TBS).
2. Incubate for 30 minutes at room temperature.
3. Rinse with TBS and incubate in TBS for 5 minutes.

F. *Preparation of chromogen (diaminobenzidine tetrahydrochloride, DAB) to fill small staining dish*
1. 6 g of ammonium acetate; 20 ml of distilled water; pH to 5.5 with concentrated citric acid.
2. Add 88 mg DAB (Sigma).
3. Cover container with foil.
4. Stir slowly for 30 minutes.
5. Filter through Whatman No. 1 filter.
6. Add 200 µl hydrogen peroxide.
7. *Note*: Add a little DAB to some of your tertiary antibody-PAP to check for brown reaction product.

G. *PAP-DAB reaction*
1. Incubate in filtered DAB until appropriate cells are stained brown; control section incubated with preimmune serum remains unstained.
2. Stop reaction in distilled water.

H. *Dehydrate*
1. 20 dips each: 70%, 95 (2×), 100% (2×), air dry.

I. *Autoradiographic dipping, in dark room with safe light.*
1. Warm emulsion to 40° C.
2. Dilute 1:1 with sterile distilled water.
3. Mix with clean slide (avoid bubbles).
4. Dip slides slowly and smoothly and stand them upright in test tube racks.
5. Store in light-tight warm, humid chamber for 2 hours, and then in black plastic slide boxes at 4° C for the appropriate times.

J. *Developing autoradiographs*
1. Prepare Kodak D-19 and fixer according to package directions (store unused portion in brown bottles at 4° C).
2. Cool developer, stop bath (distilled water), and fixer to 16° C.
3. Develop slides for 1 minute, stop for 30 seconds, and fix for 3 minutes.

K. *Staining*
1. Put slides in Meyer's hematoxylin for 8 minutes, rinse for 10 minutes in running tap water.
2. Dehydrate through alcohols and xylenes and coverslip.

Histological Staining

After *in situ* hybridization, we stain with Harris's hemotoxylin and eosin y, but since the red eosin stain decreases the signal from immunoreactive cells, Meyer's hemotoxylin is routinely used to stain sections after immunohistochemistry. Alternatively, brain tissue can be stained with Lapham's stain to intensify the immunohistochemical staining (Plates 4b and 4c). A number of cell-type-specific stains—for example, the silver stains for glial subsets— are described by Vacca (1985). Where these are compatible with *in situ* hybridization, using either radiolabeled or biotinylated nucleotide probes, they are also evidence of mRNA specific hybridization. We have colocalized GFAP immunoreactive product in astrocytes on the same tissue sections with silver-stained microglia, and Haase et al. (1985) have shown that mRNAs hybridized to two different probes can be detected in one cell. Thus, a number of mRNAs and their translation products might be detected simultaneously in a cell type. This is particularly exciting for neuroscientists because not only is there cell-type heterogeneity based on morphologic and neurotransmitter criteria, but subpopulations within a type may now be categorized according to population-specific neuropeptides. Moreover, with quantitative *in situ* hybridization, the functional relevance and regulation of these cell-type-specific gene products can be studied.

QUANTITATION OF THE LEVEL OF A SPECIFIC mRNA RELATIVE TO THE LEVEL OF TOTAL POLY-A mRNA

Problems inherent to autoradiographic grain counting per unit area or per cell, particularly in the nervous system, include cell size—the volume of cells of different types must be calculated and a standard area for counting over each area established; differences in experimental, disease, or agonal states or developmental stages—the level of a given mRNA may not be different relative to the level of total mRNA in a given state or stage; and differences in tissue preservation and processing—the availability of the mRNA for hybridization must be established for each method of preservation and processing. We circumvent these problems by hybridizing adjacent sections with radiolabeled poly-U and calculating the following:

$$\frac{\dfrac{\text{Average no. grains}}{\text{cell}} \text{Specific} - \dfrac{\text{Average no. grains}}{\text{cell}} \text{Irrelevant}}{\dfrac{\text{Average no. grains}}{\text{cell}} \text{Poly-U} - \dfrac{\text{Average no. grains}}{\text{cell}} \begin{array}{c}\text{Poly-U}\\ \text{after}\\ \text{RNase}\end{array}}$$

This formulation from reproducible, low-background hybridizations, together with negative controls without grains and positive controls when possible, allows quantitation of the amounts of a specific mRNA relative to the total poly-A mRNA in the same cell type. The only assumption necessary is that the radiolabeled poly-U and specific complementary nucleotide sequence have equal access to the mRNA in the cell. With that given, the ratio may be calculated without regard to differences in penetration, loss of RNA associated with differences in age, postmortem treatment, fixation, permeabilization treatment, and hybridization conditions. For example, the hybridization conditions for the poly-U and specific complementary sequence may be different, and the hybridization need not go to completion in either for the ratio to be calculated, since each of these factors drops out in the comparison of the ratios between two cell types, two sections from different preparations, or two sections with different degrees of permeability.

SUMMARY

A variety of fixation strategies, permeabilization protocols, synthetic and native probes, radionuclides, probe mixtures, and stringencies have now been compared, suggesting that *in situ* hybridization and immunohistochemical techniques, with their widespread applications, can be used to analyze brain cell function at the molecular level. Particularly in the brain, with its heterogeneous cell populations, many of which may be identified only by techniques detailed in this volume, the ability to analyze function in individual cells greatly increases our potential for studying how brain function is regulated during development and in health and disease.

ACKNOWLEDGMENTS

I would like to thank Lanya Lonsberry, Michael Alejos, Riki Cox, Nancy Tyler, and Marcelle Morrison for technical and other assistance with this work, which was supported in part by NIH grants AI 14336, HD 14886, and AG 05537.

REFERENCES

Angerer, L.M. and Angerer, R.C. Detection of poly A^+ RNA in sea urchin eggs and embryos by quantitative *in situ* hybridization. *Nucleic Acids Res.* 9:2819–2840, 1981.

Beckstead, J.H. and Bainton, D.F. Enzyme histochemistry on bone marrow biopsies. *Blood* 55:386–394, 1980.

Brahic, M. and Haase, A.T. Detection of viral sequences of low reiteration frequency by *in situ* hybridization. *Proc. Natl. Acad. Sci. USA* 75:6125–6129, 1978.

Capco, D.G. and Jeffrey, W.R. Differential distribution of poly(A)-containing RNA in the embryonic cells of *Oncopeltus fasciatus*. *Dev. Biol.* 67:137–151, 1978.

Denhardt, D.T. A membrane filter technique for the detection of complementary DNA. *Biochem. Biophys. Res. Commun.* 23:641–646, 1966.

Griffin, W.S.T., Alejos, M.A., and Morrison, M.R. Brain protein and messenger RNA identification in the same cell. *Brain Res. Bull.* 10:597–601, 1983.

Griffin, W.S.T., Alejos, M.A., Cox, E.J., and Morrison, M.R. The differential distribution of beta tubulin mRNAs in individual mammalian brain cells. *J. Cell Biochem.* 27:205–214, 1985.

Griffin, W.S.T. and Morrison, M.R. *In situ* hybridization—visualization and quantitation of genetic expression in mammalian brain. *Peptides* 6 (Suppl. 2):89–96, 1985.

Haase, A.T., Walker, D., Stowring, L., Ventura, P., Geballe, A., and Blum, H. Detection of two viral genomes in single cells by double-label hybridization *in situ* and color microradioautography. *Science* 227:189–192, 1985.

Harrison, P.R., Conkie, D., Affara, N., and Paul, J. *In situ* localization of globin messenger RNA formation. I. During mouse fetal liver development. *J. Cell Biol.* 63:402–413, 1974.

Jones, K.W. In *New Techniques in Biophysics and Cell Biology 1*. R.H. Pain and B.J. Smith (Eds.). London: John Wiley & Sons, 1973, pp. 29–66.

Manuelidis, L. and Ward, D.C. Chromosomal and nuclear distribution of the Hind III 1.9-kb human DNA repeat segment. *Chromosoma* (Berl) 91:28–38, 1984.

Singer, R.H. and Ward, D.C. Actin gene expression visualized in chicken muscle tissue culture by using *in situ* hybridization with a biotinated nucleotide analog. *Proc. Natl. Acad. Sci. USA* 79:7331–7335, 1982.

Steffanini, M., DeMartino, C., and Zamboni, L. Fixation of ejaculated spermatozoa for electron microscopy. *Nature* 216:172–173, 1967.

Sternberger, L.A. In *Immunocytochemistry,* 2nd Ed. New York: John Wiley & Sons, 1979, p. 24.

Vacca, L.L. *Laboratory Manual of Histochemistry* New York: Raven Press, 1985.

6

Studies of Neuropeptide Gene Expression in Brain and Pituitary

Beth S. Schachter

The long-term objective of our studies is to evaluate the effects of gonadal steroids on neuropeptide gene expression in hypothalamic neurons in the rat brain. Our interest in addressing this issue stems from observations that gonadal steroids (estrogen and progesterone) act, at the level of the hypothalamus, to regulate many aspects of female sexual and reproductive function (Pfaff, 1980). The focus on neuropeptide genes derives from observations that alterations in the levels of some neuropeptides have been correlated with changes in sexual behavior and reproductive function (Everard et al., 1980; Petraglia et al., 1985; Cholst et al., 1983; Bridges and Ronsheim, 1983; Wardlaw and Frantz, 1983; Weisner and Moss, 1982).

The rationale for asking whether or not gonadal steroids regulate the expression of specific neuropeptide mRNA levels derives from studies on the mechanisms of action of gonadal steroids on the synthesis of other proteins or peptides; in numerous cases steroid hormones have been shown to alter the amount of a protein through a change in the steady-state level of the mRNA that encodes it (Yamamoto, 1985).

At the outset of this project, the products of two genes were studied—namely, the mRNA for proopiomelanocortin (POMC; the precursor of endorphin, adrenocorticotropic hormone, and melanocyte-stimulating hormone) and mRNA for prolactin (PRL). POMC gene expression in the hypothalamus had already been documented (Liotta et al., 1979; Civelli et al., 1982; Gee et al., 1983), and there was a growing body of evidence implicating various POMC peptides in several aspects of sexual and reproductive function (Everard et al., 1980; Petraglia et al., 1985; Cholst et al., 1983; Bridges and Ronsheim, 1983; Wardlaw and Frantz, 1983; Weisner and Moss, 1982; Allen et al., 1985). Moreover, studies of RNA extracted from rat hypothalami revealed that ovariectomized animals had 30 to 40% more POMC mRNA than ovariectomized/estrogen replaced rats (Wilcox and Roberts, 1986).

Immunocytochemical studies had also detected a prolactinlike substance in rat hypothalamic neurons (Fuxe et al., 1977; Toubeau et al., 1979; Harlan et al., 1983), which projected to the midbrain (Harlan et al., 1983). In addition, studies by Harlan and coworkers showed that prolactin (purified from rat pituitaries) could potentiate some aspects of mating behavior when infused directly into the midbrains of female rats.

There are several compelling reasons for using *in situ* hybridization to assess the effects of gonadal steroids on POMC and PRL gene expression in the brain: (1) The basal hypothalamus is adjacent to the pituitary gland, an exceedingly rich source of both POMC (Civelli et al., 1982; Schachter et al., 1981) and PRL (Cooke et al., 1980) mRNAs. Although care can be taken to minimize pituitary contamination during isolation of the hypothalamus for *in vitro* RNA analyses, *in situ* localization of these mRNAs within brain cells would conclusively demonstrate intraneural expression of these genes. (2) Accurate quantitation of neuropeptides by radioimmunoassays of tissue extracts or immunocytochemical staining of tissue sections is fraught with complications on several levels. For example, posttranslational modifications of these substances may result in changes in antigenicity. Also, transport of newly synthesized neuropeptide precursors removes the substances from their sites of syntheses, frequently to many different locales in the brain. *In situ* hybridization thus provides a potentially powerful tool for measuring the steady-state neuropeptide mRNA levels, and therefore provides information on the synthetic capacity for this substance in individual cells. (3) There is increasing evidence that only a fraction of the hypothalamic POMC- and immunoreactive PRL-containing neurons concentrate gonadal steroids (Morrell et al., 1985; Shivers et al., submitted), and therefore might directly respond to these stimuli. *In situ* hybridization may reveal whether or not the estrogen-mediated response occurs in only a subset of hypothalamic neurons that produce a given neuropeptide, and whether or not the response is in the subset that contains the estrogen receptors.

TECHNICAL OPTIMIZATION

Tissue Treatment

In setting up the *in situ* hybridization assay, we took advantage of the fact that POMC and PRL, and their corresponding mRNAs are produced in large quantities in the pituitary gland (Schachter et al., 1981; Cooke et al., 1980). Thus, the histological distribution of these mRNAs can be rapidly identified by treating pituitary sections with ^{32}P-labeled (Rigby et al., 1977) nucleic acid probes (Cooke et al., 1980; Eberwine and Roberts, 1983; Seeburg et al., 1977) and detecting the corresponding hybridization signals by x-ray film autoradiography. Figure 6-1 shows an x-ray autoradiogram from duplicate microscope slides containing three pituitary sections each. Equal amounts of the indicated ^{32}P-labeled cloned DNA probes were applied to each section. (Note that a probe for the mRNA for growth hormone, another major pituitary peptide hormone, was also used.) Following hybridization and extensive washing, the slides were exposed overnight to x-ray film. The results correspond with the known distribution and relative abundance of these three mRNAs.

Our initial protocol employed a tissue fixation that yielded good cellular morphology and detectability of peptide hormones in pituitary and brain (Harlan et al., 1983). Tissue preparation involved perfusion of the rat, first with phosphate-buffered saline (PBS) alone, then with 3% paraformaldehyde in PBS. Pituitaries and hypothalami were removed from decapitated animals, they were further fixed by soaking in the paraformaldehyde solution, cryoprotected with sucrose, frozen, and cut into 6 to 10-μm sections, which were mounted onto subbed microscope slides (Harlan et al., 1983).

Subsequent tissue treatment was essentially that described by Brahic and

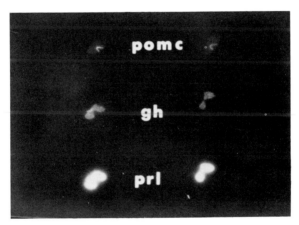

Figure 6-1 Histologic distribution and relative abundance of POMC, growth hormone (GH), and PRL mRNAs in adult female rat pituitary. Eight-micron sections through a 4% paraformaldehyde perfusion-fixed pituitary were mounted three sections per slide. POMC, GH, or PRL cDNAs were ^{32}P-labeled by nick translation (Rigby et al., 1977) to specific activities of ~10^8 cpm/μg. Twenty-microliter aliquots of hybridization buffer, containing 50 to 100 ng of labeled DNA was applied to each section. Following overnight incubation at room temperature, slides were washed to remove unreacted probe, air dried, then exposed to Kodak XAR5 film for 16 hours. (See Protocol, p. 122, for procedural details.) The x-ray film shows duplicate sides: top sections, POMC; midsections, GH; and bottom sections, PRL.

The results are consistent with the known distribution and relative abundance of POMC, GH, and PRL peptides and their corresponding mRNAs in the adult female rat pituitary: the POMC gene is expressed at a high level in all cells of the intermediate lobe (the narrow, curved band in the top sections) and in ~5% of the cells in the anterior lobe. GH gene expression is seen in ~20% of the cells in the anterior lobe cells. None of these genes is expressed in the posterior (neural) lobe, as confirmed here by the absence of signal over the semicircle of tissue surrounded by the intermediate lobe.

Haase (1978). This procedure provided adequate signal detection of the peptide hormone mRNAs in the pituitary, as assessed by x-ray film autoradiography of ^{32}P-probes, or nuclear emulsion autoradiography of ^{3}H-probes. No specific hybridization was detected in the hypothalamus, however, even after emulsion exposures of several months. Equally troubling was the finding that perfusion fixation preserved a substance in the brain that caused positive chemography. This chemographic artifact was detected in a series of negative control experiments, including omission of radioactive probe from the hybridization buffer. As shown in a coronal section through the hypothalamus (Plate 5a), the chemography resulted in silver grain reduction, which had a bilaterally symmetric distribution in the medial basal region; this pattern is very similar to the expected distribution of POMC mRNA-containing neurons.

Positive chemography was seen in sections of perfusion-fixed animals only when they were carried through the hybridization protocol. Sections that were untreated before emulsion coating were free of reduced grains even after prolonged exposure to the emulsion. Therefore, we hypothesized that the method of fixation preserved a substance in the tissue that was chemically altered during the hybridization or washings steps, and could then interact with the silver.

Hence, other methods of tissue fixation were evaluated. Tissue was fresh-frozen in OCT, cut, mounted onto subbed slides, air dried briefly at room temperature, then fixed in buffered paraformaldehyde, ethanol/acetic acid (3:1), Carnoy's fixative, or Bouin's fixative (Lillie and Fullmer, 1976). Paraformaldehyde fixative was removed by rinsing in PBS, then in water. These slides were dehydrated through alcohols. Other fixatives were removed with alcohol. Hypothalamic sections treated with each of the fixatives were processed through the hybridization protocol in the absence of radioactive probe, coated with emulsion, stored for several days, photodeveloped, and examined. All the samples of fresh-frozen/postfixed tissue were found to be free of the positive chemographic artifact.

Pituitary sections treated with each of the fixatives were compared in the hybridization assay using ^{32}P-labeled POMC or PRL DNA probes. In addition, the effects of deproteinization, by treatment with HCL and/or proteases, and increasing permeability by nonionic detergent treatment were evaluated. It was repeatedly found that postfixation with 3 to 4% paraformaldehyde for 5 minutes with no additional treatment gave the highest specific hybridization signal. Also, the addition of diethyl pyrocarbonate (DEP; 0.02% v/v) to the paraformaldehyde solution just prior to use (Gee and Roberts, 1983) increased the specific hybridization signal, possibly by inactivating ribonucleases in the reagents or the tissue.

Details of the protocol we now use are presented on p. 122 under Protocol. It is noteworthy that this protocol involves a minimum of manipulations before hybridization. The fact that several other workers have reported that deproteinization increases signal strength (Brahic and Haase, 1978; Gee and Roberts, 1983; Cox et al., 1984), while we observed the converse, can perhaps be reconciled by considering that the benefits of deproteinization were most often seen when tissue was subjected to extensive fixation rather than to a limited one such as that reported here. Extensive fixation may block accessability to mRNA, which is then relieved by acid or protease treatment, or both. In contrast, limited fixation followed by deproteinization may cause significant losses of mRNA during the hybridization and washing (Singer, personal communication).

Hypothalamic sections, postfixed with paraformaldehyde, were then tested, and cells that reacted with the POMC DNA probe were readily detected, as can be seen in Plates 5b through 5d. No comparable hybridization signal was seen with the PRL DNA probe.

Hybridization Probes

Initial experiments used double-stranded DNA probes. Plasmids containing DNA sequences complementary to rat pituitary POMC (Eberwine and Roberts, 1983) and PRL (Cooke et al., 1980) mRNA were digested with restriction enzymes, and the rat DNA sequences were fractionated away from plasmid DNA by gel electrophoresis (Maniatis et al., 1982), then radiolabeled by nick-translation (Rigby et al., 1977). The conditions of the labeling reaction were adjusted to generate DNA fragment lengths of about 50 to 200 nucleotides.

It is worth noting that the initial PRL DNA fragment used was the entire 823 base pair cDNA insert, including GC tails of undetermined length (Cooke et al., 1980). This probe gave unacceptably high backgrounds in both *in situ* and RNA gel blot hybridizations. The problem was overcome by using an internal 640-base-pair fragment of this DNA sequence, which is free of GC homopolymeric regions.

Comparisons were made of the signal strength and signal-to-noise ratio obtained from ^{3}H-, ^{32}P-, and ^{35}S-labeled probes. Plates 5b through 5d show high magnification views through hypothalamic sections of a rat brain probed with POMC DNA labeled with each of the three isotopes. As expected, ^{3}H provided the best cellular localization of label and gave very low backgrounds (see Plate 5b). ^{32}P, which gives reduced silver grains much further from its source than ^{3}H, was found to be adequate for cellular localization of POMC-containing neurons in the brain because these cells are large and relatively far apart (see Plate 5c). ^{35}S gave good cellular localization, and allowed for signal detection of POMC mRNA in brain much sooner than the ^{3}H-labeled probe (see Plate 5d). However, the background generated by thiol-labeled nucleotides was consistently higher than with either ^{3}H or ^{32}P. The most acceptable signal-to-noise ratio obtained with ^{35}S-labeled probes (e.g., that is shown in Plate 5d) resulted from experiments in which unlabeled alpha-thiol nucleotide (5 μM; New England Nuclear) and beta-mercaptoethanol (14.7 mM) were included in the hybridization buffer, mercaptoethanol and sodium thiosulfate (1%) were included in the wash buffer, and washing times were extended to 2 days rather than 1 (Shivers et al., 1986). Taken together, these results led us to conclude that ^{3}H was the isotope of choice for quantitative analysis at the cellular level.

Because specific hybridization of the PRL DNA probe was not seen, we made further effort to increase the signal strength and signal-to-noise ratio. To this end, the PRL and POMC DNAs were subcloned into plasmids which can serve as templates for generating single-stranded RNA probes (SP; Promega). These so-called riboprobes, when tested on pituitary sections, most often gave lower backgrounds than their double-stranded DNA counterparts. The hybridization and washing conditions were essentially those described by Cox et al. (1984), with minor modifications described in Protocol. No noticeable increase in signal intensities was seen with riboprobes.

The PRL riboprobe, while clearly detecting the pituitary mRNA, showed no specific localization in brain sections. This result, in conjunction with other findings in our laboratories suggested that the original observation of immunoreactive (IR) PRL in the hypothalamus should itself be questioned. Specifically, we considered the possibility that anti-PRL antisera cross-reacted with substances other than prolactin in brain tissue. Subsequent studies did show that IR-PRL was not detected in hypothalamic neurons when the PRL antiserum was preabsorbed with the amino-terminal glycopeptide fragment of POMC. A detailed discussion of this issue is presented elsewhere (Harlan et al., submitted). These results demonstrate one of the important uses of *in situ* mRNA hybridization—namely, as a means of verifying (or disputing) the true specificity of antisera to the corresponding protein.

COMBINING IMMUNOCYTOCHEMISTRY AND *IN SITU* HYBRIDIZATION ON THE SAME TISSUE SECTION

The effects of steroid hormones on neuropeptide gene expression could be evaluated readily by identifying steroid-responsive cells and neuropeptide mRNAs on the same tissue sections. Such information would be obtainable by combining immunocytochemical detection of a steroid receptor with quantitative *in situ* hybridization for a specific neuropeptide mRNA.

We therefore evaluated conditions for localizing protein, by immunocytochemistry, and mRNAs, by *in situ* hybridization, on the same tissue section (Shivers et al., 1986). Several other workers have reported successful use of this combination of techniques. For example, viral antigens and viral mRNA have been detected in the same sections (Brahic et al., 1984), and immunoactive corticotropin-releasing factor (CRF) and vasopressin mRNA have been colocalized (Wolfson et al., 1985).

Once again we used the pituitary as a model system for evaluating technical aspects of the combination assay. Nucleic acid probes for POMC and PRL, and antibodies for ACTH (the midportion of POMC) and PRL were tested. To best preserve antigenicity, rats were perfused with buffered paraformaldehyde (Harlan et al., 1983). One set of frozen sections was subjected to immunocytochemical treatment followed by *in situ* hybridization while another set was given the reverse order of treatments.

Primary antisera used in this experiment were rabbit antirat PRL (National Hormone and Pituitary Program; prepared by A. Parlow) rabbit anti-ACTH (AC-6; gift of R. Benoit) or rabbit anti-beta-endorphin (beta-Endo-2; gift of R. Benoit). Immunocytochemical detection used the avidin-biotinylated horseradish peroxidase method (ABC) with reagents from Vector, used according to suppliers' recommendations. *In situ* hybridization used ^{3}H-labeled nick-translated PRL and POMC DNA probes, as described in Protocol. Employing our standard protocols in either reaction sequence (immunocytochemistry followed by *in situ* hybridization or the reverse) rendered the second substrate completely undetectable.

We hypothesized that when the immunocytochemical detection is done first, ribonucleases in the antisera may hydrolyze a significant amount of mRNA and thus prohibit successful subsequent *in situ* hybridization. We therefore tested the effects of including each of several ribonuclease inhibitors in the primary and secondary antisera. Vanadyl ribonucleoside (BRL) was used at 15 mM, RNasin (gift of Dr. P. Blackburn) at 180 U/ml, and diethyl pyrocarbonate (DEP) at 0.04% v/v.

DEP, but not the other two ribonuclease inhibitors, successfully protected the RNA during incubation of tissue with antisera. As seen in Plates 6a through 6c, various combinations of antiserum and nucleic acid probe were used to detect antigen and mRNA either in the same cell population (Plate 6a: anti-PRL antiserum and PRL cDNA probe) or in nonoverlapping cell types (Plate 6b: anti-PRL and POMC cDNA probe; Plate 6c: anti-ACTH and PRL cDNA).

This sequential detection method also successfully revealed antigen and mRNA in fresh-frozen/paraformaldehyde postfixed pituitary, as can be seen in Plate 6d. As noted previously, there is some loss of cellular morphology and peptide immunoreactivity inherent in this method of tissue preparation. Nonetheless, Plate 6d shows that PRL and PRL mRNA are colocalized within a subset of anterior pituitary cells, and neither product is detected elsewhere in the pituitary.

QUANTITATIVE *IN SITU* HYBRIDIZATION

Before examining at the cellular level the effects of gonadal steroids on neuropeptide gene expression, it is first necessary to verify that the *in situ* hybridization technique gives reproducible results on several animals in a single treatment group. To this end, a pilot experiment was run in which POMC gene expression was evaluated in hypothalamic sections of several ovariectomized rats 1 week after estrogen replacement. Fresh-frozen/paraformaldehyde postfixed tissue was cut into 10-μm sections, and mounted three per slide. Samples were hybridized with ^{3}H-POMC cRNA; three probe concentrations (0.06, 0.13, and 0.175 ng μl; 20 μl per section) were used on each slide. Emulsion autoradiography using undiluted NTB2 emulsion was carried out for 32 days. All the samples were processed in parallel at all stages of the experiment.

To chart the labeled cells, an image of each section was projected under a microscope lens onto a large sheet of paper. Tissue landmarks were traced onto the paper. Each section was then viewed under the light microscope, and the position of each labeled cell and the number of grains per labeled cell were recorded onto each map.

In addition, the background level of silver grains was measured on each section, in the dorsal region of the tissue, where no specific POMC mRNA is expected. The average number of grains found over 10 medium-to-large cells in this region was taken to be the nonspecific background for each section; this value was subtracted from the grain count of all specifically labeled cells.

The sections probed with the highest concentration of cRNA had many well-labeled cells. It is important to note that the 32-day exposure time did not appear to saturate the emulsion; such saturation would yield many large, unresolvable silver grains that could not be counted accurately.

Limited analysis of sections probed with the two highest probe concentrations showed a proportional increase in number of grains per labeled cell. These results indicate that before a large experiment is set up to compare treatment groups, it will be necessary to test even higher probe concentrations, to verify that the POMC mRNA hybridization has been driven to completion. Failure to saturate the mRNA could result in an attenuation of signal differences among the sample treatment groups.

Table 6-1 shows the results of the data analyzed to date. Values in this table were obtained from tissue sections that received the highest probe concentration. At least two sections from each of three brains were analyzed. Cells were scored as specifically labeled if they had ≥ 3 times the mean background grain number. Several important features are revealed here: (1)

Table 6-1   Comparison of POMC *in situ* hybridization signal in hypothalami of endocrinologically equivalent rats

Brain no.	Slide no.	Background grains/cell[a]	Labeled cells	Grains/cell[b]
1	52	1.7	37	15.6
	49	2.5	25	10.0
	58	0.7	34	11.5
2	55	1.4	42	11.9
	59	1.2	16	9.9
3	27	1.7	27	13.2
	32	2.1	17	12.6

[a]Values represent the mean number of silver grains found over 10 medium to large cells in the dorsal region of the hypothalamus on each section.

[b]Values represent the mean grain count per labeled cell after the slide background is subtracted.

the mean background grain count is quite low (range of means = 0.7–2.5 grains per cell) and there was no systematic difference between brains; (2) the range of numbers of labeled cells per section (range = 16–42) also showed no systematic difference among the brains; and (3) the background was sufficiently low in all sections to allow for easy identification of numerous labeled cells.

We therefore conclude that the method is sufficiently sensitive and reproducible to detect small, gonadal steroid-mediated changes in either the number of cells containing POMC mRNA or the amount of mRNA per cell. Studies of this sort are now underway.

SUMMARY AND CONCLUSION

The aim of this project has been to optimize the *in situ* hybridization technique for quantitative studies on the effects of gonadal steroids on neuropeptide gene expression in the rat brain. The reasons for using *in situ* hybridization are threefold: (1) the technique avoids the problem of contamination of RNA from other tissues that may be an enriched source of the mRNA of interest; (2) it avoids the problems encountered in quantitative radioimmunoassays (RIAs) of tissue extracts and quantitative immunocytochemistry; and (3) it provides a means of anatomically mapping hormone-responsive cells within small regions of the brain that might contain a population of heterogeneously responsive cells.

Our initial attempts at localizing POMC and PRL mRNAs in brain sections used tissue fixation techniques optimized for immunocytochemical detection of neuropeptides. The original hybridization procedure was adapted from a protocol optimized for detecting very long viral RNA sequences in brain tissue.

Surprisingly, we found that in combining these protocols, a substance was preserved in the brain (but not pituitary) tissue, which gave a positive chemographic reaction with the nuclear emulsion used to detect the hybridization signal. The chemographic silver grain reduction occurred in the very same area of tissue in which hybridization was to be detected. Therefore, alternative methods of tissue fixation and subsequent treatment were investigated to find an appropriate combination that gave readily detectable signal, high signal-to-noise ratio, and no chemographic artifacts. Of the various combinations of protocols investigated, the best one involved cutting frozen, unfixed tissue sections, which were then briefly postfixed on microscope slides with 3 to 4% buffered paraformaldehyde containing the ribonuclease inhibitor DEP. Subsequent treatments, such as deproteinization or detergent permeabilization were found to be without benefit.

Comparisons were made of the hybridization signal strength and signal-to-noise ratio using nucleic acid probes labeled with ^{3}H, ^{32}P, or ^{35}S. Tritium-labeled probes were found to be the best choice for quantitative studies at the cellular level. In addition, single-stranded cRNA probes, compared with their double-stranded DNA counterparts, were found to give a higher specific signal-to-noise ratio.

Because the proposed research on the effects of gonadal steroids will be most informative when immunocytochemical detection of steroid receptors can be combined with *in situ* hybridization of neuropeptide mRNAs, preliminary studies were carried out to determine the technical requirement for such colocalization. These investigations have thus far focused exclusively on the readily detectable signals in the pituitary gland, using antisera to POMC-derived peptides and PRL, and the corresponding nucleic acid probes. This work revealed that colocalization of the products was readily achieved by sequential immunocytochemistry and *in situ* hybridization if the initial antisera reactions were done in the presence of DEP. Colocalization of peptide and mRNA was demonstrated on both tissue sections from perfusion-fixed animals and on fresh-frozen/postfixed sections. The latter method of tissue fixation not only avoids problems of chemography but also may be preferable for maintaining antigenicity of the estrogen receptor (Sar and Parik, 1984) in our future studies.

Finally, it was necessary to verify that the *in situ* hybridization protocol is sufficiently reproducible to detect small changes in neuropeptide gene expression. A test was set up to compare the hybridization signal of the POMC probe on brain sections of several animals who were all in equivalent endocrine states. The results are encouraging; backgrounds were consistently low after a 1-month exposure, and the number of labeled cells per section and grains per cell did not vary more among the animals than they did among different sections of each animal.

Therefore, it is concluded that the technique of *in situ* hybridization should provide a powerful tool for future studies on the regulation of neuropeptide gene expression.

ACKNOWLEDGMENTS

This work is the result of a most exciting and instructive collaboration with Dr. Brenda Shivers. I am pleased to also acknowledge the scientific contributions of Drs. Richard Harlan and Donald Pfaff, the technical support of Wing Chiu and Jane Doeblin, and the secretarial assistance of Melanie Ashley-Walker. This study has been funded in part by NIH grant HD18110.

REFERENCES

Allen, D.L., Renner, K.J., and Luine, V.N. Naltrexone facilitation of sexual receptivity in the rat. *Hormo. Behav.* 19:98, 1985.

Brahic, M. and Haase, A.T. Detection of viral sequences of low reiteration frequency by *in situ* hybridization. *PNAS* 75:6125, 1978.

Brahic, M., Haase, A.T., and Cash, E. Simultaneous *in situ* detection of viral RNA and antigens. *PNAS* 81:5445, 1984.

Bridges, R.S. and Ronsheim, P.M. Changes in beta-endorphin concentrations in the medial preoptic area during pregnancy in the rat. *Soc. Neurosci. Abstr.* 233(13):799, 1983.

Cholst, I.N., Wardlaw, S.L., and Frantz, A.G. Pseudo-pregnancy increases brain beta-endorphin. *Endocrinol. Soc. Abstr.* 1167:372, 1983.

Civelli, O., Birnberg, N., and Herbert, E. Detection and quantitation of pro-opiomelanocortin mRNA in pituitary and brain tissue from different species. *J. Biol. Chem.* 257:6783, 1982.

Cooke, N.E., Coit, D., Weiner, R.I., Baxter, J.D., and Martial, J.A. Structure of cloned DNA complementary to rat pituitary messenger RNA. *J. Biol. Chem.* 255:6502, 1980.

Cox, K.H., DeLeon, D.V., Angerer, L.M., and Angerer, R.C. Detection of mRNAs in sea urchin embryos by *in situ* hybridization using asymmetric RNA probes. *Dev. Biol.* 101:485, 1984.

Eberwine, J.H. and Roberts, J.L. Analysis of POMC gene structure and function. *DNA* 2:1–8, 1983.

Everard, D., Wilson, C.A., and Thody, A.J. Effect of alpha-melanocyte-stimulating hormone and related peptides on lordosis behavior in female rats. *J. Endocrinol.* 85:2P, 1980.

Fuxe, K., Hokfelt, T., Enroth, P., Gustafsson, J.-A., and Skett, P. Prolactin-like immunoreactivity: Localization in nerve terminals of rat hypothalamus. *Science* 196:899, 1977.

Gee, C.E., Chen, C.L.C., Roberts, J.L., Thompson, R., and Watson, S.J. Identification of pro-opiomelanocortin neurons in rat hypothalamus by *in situ* cDNA-mRNA hybridization. *Nature* 306:374, 1983.

Gee, C. and Roberts, J.L. *In situ* hybridization histochemistry: A technique for the study of gene expression in single cells. *DNA* 2:155, 1983.

Harlan, R.E., Shivers, B.D., Schachter, B.S., Kaplove, K.A., and Pfaff, D.W. Distribution and partial characterization of immunoreactive prolactin in the rat brain. Manuscript submitted.

Harlan, R.E., Shivers, B.D., and Pfaff, D.W. Midbrain microinfusions of prolactin increase the estrogen-dependent behavior, lordosis. *Science* 219:1451, 1983.

Lillie, R.D. and Fullmer, H.M. *Histopathologic, Technical and Practical Histochemistry.* New York:McGraw-Hill, 1976.

Liotta, A., Gildersleeve, D., Brownstein, M.J., and Krieger, D.T. Evidence for brain synthesis of immunoreactive ACTH and beta-endorphin-like activity. *PNAS* 76:1448, 1979.

Maniatis, T., Fritsch, E.F., and Sambrook, J. *Molecular Cloning: A Laboratory Manual.* New York:Cold Spring Harbor Laboratory, 1982.

Morrell, J.I., McGinty, J., and Pfaff, D.W. A subset of beta-endorphin- or dynorphin-containing neurons in the medial basal hypothalamus accumulates estrogen. *Neuroendocrinol.* 41:417, 1985.

Petraglia, F., Baraldi, M., Giarre, G., Facchinetti, Santi, M., Volpe, A., and Genazzani, A.R. Opioid peptides of the pituitary and hypothalamus: Changes in pregnant and lactating rats. *J. Endocrinol.* 105:239, 1985.

Pfaff, D.W. *Estrogens and Brain Function.* New York: Springer-Verlag, 1980.

Rigby, P.W.J., Dieckmann, M., Rhodes, C., and Berg, P. Labeling deoxy-ribonucleic acid to high specific activity *in vitro* by nick translation with DNA polymerase I. *J. Mol. Biol.* 113:237, 1977.

Sar, M. and Parik, I. Immunohistochemical localization of estradiol receptor in rat brain with monoclonal antibodies to calf cytosolic receptor. *Inter. Cong. Endocrinol. Abstr.* No. 2415, 1984.

Schachter, B.S., Johnson, L.K., Baxter, J.D., and Roberts, J.L. Differential regulation by glucocorticoids of pro-opiomelanocortin mRNA levels in the anterior and intermediate lobes of rat pituitary. *Endocrinol.* 110:1442, 1981.

Seeburg, P.H., Shine, J., Martial, J.A., Baxter, J.D., and Goodman, H.M. Nucleotide sequence and amplification in bacteria of structural gene for rat growth hormone. *Nature* 270:486, 1977.

Shivers, B.D., Harlan, R.E., and Pfaff, D.W. Manuscript submitted.

Shivers, B.D., Schachter, B.S., and Pfaff, D.W. *In situ* hybridization for the study of gene expression in the brain. In *Methods in Enzymology* 124:497–510, 1986.

Shivers, B.D., Harlan, R.E., Pfaff, D.W., and Schachter, B.S. Co-localization of peptide hormones and peptide hormone mRNAs in the same tissue sections of rat pituitary. *J. Histochem. Cytochem.* 34:39, 1986.

Toubeau, G., Desclin, J., Parmentier, M., and Pastecls, J.L. Cellular localization of a prolactin-like antigen in the rat brain. *J. Endocrinol.* 83:261, 1979.

Wardlaw, S.L. and Frantz, A.G. Brain beta-endorphin during pregnancy, parturition and the post-partum period. *Endocrinol.* 113:1664, 1983.

Weisner, J.B. and Moss, R.L. Beta-endorphin suppression of mating behavior and plasma luteinizing hormone in the female rat. *Soc. Neurosci. Abstr.* 8:930, 1982.

Wilcox, J.N. and Roberts, J.L. Estrogen decreases rat hypothalamic pro-opiomelanocortin messenger ribonucleic acid levels. *Endocrinol.* 117:2392, 1986.

Wolfson, B., Manning, R.W., David, L.G., Arentzen, R., and Baldino, F., Jr. Co-localization of corticotropin releasing factor and vasopressin mRNA in neurons after adrenalectomy. *Nature* 315:59, 1985.

Yamamoto, K.R. Steroid receptor regulated transcription of specific genes and gene networks. *Annu. Rev. Genet.* 19:209, A. Campbell, I. Herskowitz and L.M. Sandler (Eds.), 1985.

PROTOCOL

Slide Preparation (Subbing)

1. Use precleaned, frosted, and microscope slides.
2. Soak slides 20 minutes in 0.2 N HCl in 95% ethanol.
3. Rinse three times in distilled, deionized H_2O.
4. Soak the slides for 2 minutes in 0.02% DEP-treated 0.5% melted gelatin, 0.05% chrom-alum.
5. Drain the slides, then dry them at 60° C and store in a dust-free place.

Tissue Freezing

1. Prepare a small, flat-bottomed aluminum foil boat.
2. Place a drop of OCT (Tissue TeK II) in the boat, then orient the tissue block in the boat. Fill the boat with OCT to completely cover tissue.

3. Chill a small amount of isopentane in liquid nitrogen.
4. With forceps, hold the boat in the chilled isopentane solution, being careful to keep the coolant from spilling into the boat.
5. Tissue may be covered tightly with aluminum foil and stored for several weeks at −70° C before it is sectioned.

Cutting and Fixing the Tissue

1. Cool the microtome chuck on dry ice.
2. Mount the tissue block onto the chuck with additional OCT.
3. Allow the tissue to warm in a cryostat chamber (to −20° C to −16° C for unfixed tissue).
4. Cut sections (6–10 μm) and mount them three per slide.
5. Let the tissue air dry onto the slide for 2 to 3 minutes after the third section is applied.
6. Fix tissue for 5 minutes in 3% paraformaldehyde in PBS containing 0.02% DEP.
7. Rinse 2 × 2 minutes in PBS.
8. Rinse for 1 minute in the following solutions: water, 50% ethanol, 70% ethanol, 95% ethanol, and 100% ethanol.
9. Air dry the slides, then store them at −70° C until needed.

Prehybridization, hybridization, and washing of slides probed with nick-Translated DNAs

Ingredient	Concentration in prehybridization solution	Concentration in hybridization solution
NaCl	600 mM	600 mM
Tris HCl	10 mM	10 mM
Ficoll	0.02 %	0.02 %
Polyvinyl pyrrolidone	0.02 %	0.02 %
Bovine serum albumin (Fraction V)	0.02 %	0.02 %
Na$_2$EDTA	1 mM	1 mM
Yeast total RNA	0.05 %	0.005%
Yeast tRNA	0.005%	0.005%
Herring sperm DNA (sonicated)	0.05 %	0.01 %
Sodium pyrophosphate	0.05 %	0.05 %
Dextran sulfate (8000)	0.05 %	10 %
Formamide (deionized)	50 %	50 %
Alpha-thiol NTP		0.005 mM[a]
Mercaptoethanol		14.7 mM[a]

[a]For ^{35}S probes only.

1. Transfer slides from −70° C to a room-temperature dessicator for 20 minutes.
2. Encircle each tissue section with clear nail polish (e.g., Revlon Wet'n Wild). This provides a barrier so that different probes can be used on a single slide.
3. For incubation chambers, use clear plastic utility boxes (Fisher). Place a sheet of 3 MM filter paper in each. Saturate filters with 600 mM NaCl, 50% formamide.

4. Lay out the slides in a box. Boil the prehybridization solution for 10 minutes, then chill on ice. Apply 30 to 50 μl of prehybridization solution to each section.
5. Seal the box and incubate for 1 hour at 37° C.
6. Blot off prehybridization solution.
7. Apply 30 to 50 μl hybridization solution to each section (preboiled as before and containing 1 to 5×10^4 cpm radiolabeled probe).
8. Seal the box and incubate for 48 to 72 hours at 37° C.

Washing (for DNA Probes)

1. Blot off hybridization buffer with a paper towel.
2. Wash slides 2×10 minutes in 1 to 2 liters of 0.3 M NaCl, 0.03 M sodium citrate, 0.05% sodium pyrophosphate at room temperature. (Include 14.7 mM beta-mercaptoethanol and 1% sodium thiosulfate for ^{35}S probes.)
3. Wash the slides for 24 hours in 2 to 4 liters of 0.075 M NaCl, 0.0075 M sodium citrate, 0.05% sodium pyrophosphate at 37° C. (For ^{35}S probes include mercaptoethanol and sodium thiosulfate as before, and increase the washing time to 48 hours.)
4. Remove nail polish.
5. Dehydrate slides in 70% ethanol, 300 mM ammonium acetate; 90% ethanol, 300 mM ammonium acetate, and 100% ethanol.

Washing (for Single-Stranded RNA Probes)

1. Following step 2 in the preceding section, treat tissue with 20 μg/ml RNase A in 0.5 M NaCl, 10 mM Tris, pH 8.0, 1 mM EDTA, at 37° C for 30 minutes.
2. Rinse the slides in the previously described buffer without RNase.
3. Rinse overnight, as described for DNA probes.

Emulsion Autoradiography and Staining

For maximum signal intensity, these studies have used undiluted Kodak NTB2 emulsion.

1. Melt the emulsion by heating to 37 to 40° C for 3 hours.
2. Working under a sodium vapor safe light (duplex Super Safelight; Thomas), pour emulsion in a slide mailer (Fisher; ~28-ml capacity, enough for 30 slides).
3. Place the slide mailer in a beaker of water prewarmed to 42° C on a hot plate.
4. Allow the emulsion to reequilibrate for 5 minutes.
5. Dip each slide slowly in and out of the emulsion twice. Inspect under a safe light to confirm an even coating.
6. Dry the slides overnight in a rack stored within a light-tight box in a dark cabinet.
7. Box the slides in light-tight Bakelight boxes. In each box place a few dessicant capsules (Humicaps; Driaire). Tape the boxes with electrical tape and wrap them in aluminum foil.
8. Store at 4° C until photodevelopment.
9. On the day of photodevelopment, allow the slide boxes to come to room temperature.
10. Have the developer, water, stop, and fixer at 15 to 17° C.
11. Develop in Kodak D19 developer for 3 minutes.
12. Rinse for 30 seconds in water.

13. Fix 2 × 5 minutes in Kodak fixer.
14. Rinse in several changes of water.
15. Stain the tissue in 0.5% cresyl violet in water.
16. Rinse with water, graded alcohols, and xylenes.
17. Coverslip with Permount mounting medium.

7

Anatomical Localization of mRNA: *In Situ* Hybridization of Neuropeptide Systems

Stanley J. Watson, Thomas G. Sherman,
Jeffrey E. Kelsey, Sharon Burke,
and Huda Akil

The purpose of this chapter is to present the logic and methodology associated with *in situ* hybridization in nervous and endocrine tissue. The primary goal of *in situ* hybridization is the visualization of a specific messenger RNA (mRNA) in a precise anatomical context within a particular cell or cell group. A specific mRNA codes for only one protein, through a unique sequence of nucleic acids (mRNAs are often between 600 and 2000 bases in length). A particular cell contains many species of mRNA, some coding for common proteins, such as tubulin, or various membrane and structural proteins that serve general cellular purposes, and others coding for proteins specific to that cell type. In the case of a neuron or endocrine cell that uses a peptide as its neurotransmitter, or hormone (or neuromodulator), the cell produces the mRNA coding for the precursor to that neuropeptide. For example, a vasopressin/neurophysin-producing neuron in the paraventricular region of the hypothalamus produces these peptides from the specific mRNA coding for their precursor protein. Obviously, specific mRNAs coding for receptors, second messenger proteins, ion channels, and transmitter-synthesizing enzymes are of similar major interest to neuroscientists.

In immunohistochemistry, one often visualizes a particular protein or peptide using specific antisera. By analogy, *in situ* hybridization allows the visualization of a specific mRNA species using a specific, labeled complementary nucleic acid (either DNA or RNA) strand as the indicator. (See Figure 7-1 for a graphic illustration of this logic and Figure 7-2 for examples of *in situ* hybridization.) The principal important issue here is that two DNA strands (or one DNA and one mRNA, or one mRNA and one cRNA strand) can become double-stranded (e.g., hybridized) under the proper conditions

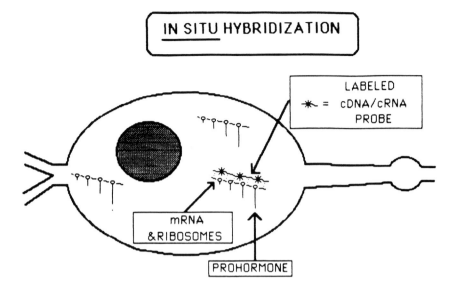

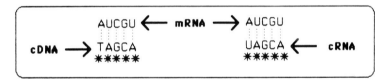

Figure 7-1

(time, temperature, salt), providing each strand can form a hydrogen bond with its potential partner (guanine and cytosine base pair, and thymine and adenine base pair). For example, if one starts with an mRNA strand of the following sequence

$$5' \ldots \text{AUGCA} \ldots 3'$$

its perfect cDNA match would be $3' \ldots$ TACGT $\ldots 5'$ (mRNA uses the pyrimidine base uracil instead of thymine), resulting double-stranded nucleic acids of the following structure

    .... AUGCA .... (mRNA)   .... AUGCA .... (mRNA)

or

    .... TACGT .... (cDNA)   .... UACGU .... (cRNA)

If such a double-stranded RNA–DNA (or RNA–RNA) hybrid is long enough (around 20 bases), it can be very stable (because of the hydrogen bonds

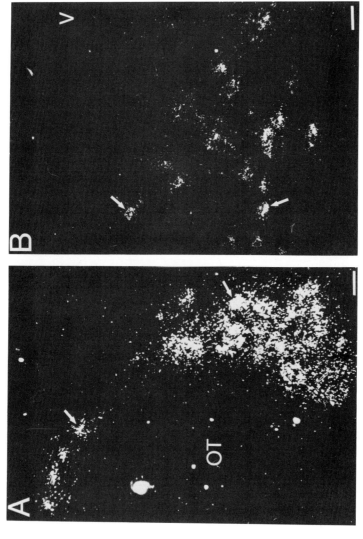

Figure 7-2 (A) Darkfield photomicrograph of rat brain supraoptic nucleus hybridized *in situ* with [³⁵S]-labeled AVP. The grains are clustered over individual cells (*arrows*) in the supraoptic nucleus. OT = optic tract; bar = 25 μm. (B) Darkfield photomicrograph of rat brain arcuate nucleus hybridized with [³⁵S]-labeled SP6 POMC riboprobe. The grains are clustered over arcuate cells (*arrows*). V = third ventricle; bar = 25 μm.

formed by each base of the pair). The requirement of complementary base matching over a long stretch of bases means that cDNA matching with mRNA (cRNA with mRNA) can be used in a highly specific fashion (analogous to an antibody "identifying" a particular protein or peptide fragment).

The *in situ* hybridization method is "complex" because it remains in its late-middle stages of development and requires the integration of classic histochemical methods with those used in the molecular genetics laboratory. In the following pages, we present a summary of the methods required for *in situ* hybridization, with a major emphasis on "monospecific" probes of a constant length, such as the synthetically produced oligonucleotides and the single-stranded RNA probes produced by the SP-6 Riboprobe system. The variety of approaches to *in situ* hybridization, as well as the diversity of problems to which it has been applied, are thoroughly discussed in the other references and in the other chapters of this volume as well as (John et al., 1977; Brahic and Haase, 1978; McWilliams and Boime, 1980; Haase et al., 1981; Pochet et al., 1981; McDougall et al., 1982; Singer and Ward, 1982; Gee et al., 1983; Gee and Roberts, 1983; Griffin et al., 1983; Hafen et al., 1983; Lynn et al., 1983; Royston and Augenlicht, 1983; Saber et al., 1983; Cox et al., 1984; Pfeifer-Ohlsson et al., 1984; Pintar et al., 1984; Scheller et al., 1984; Southern et al., 1984; Haase et al., 1985; Kornberg et al., 1985; Shaw et al., 1985; Wolfson et al., 1985; Lewis et al., 1986b; Kelsey et al., 1986).

In general our emphasis is on the methods we have adapted for the production of radiolabeled *in situ* hybridization in nervous tissue using both the oligonucleotide and the cRNA probe methods. Our secondary emphasis is the nature and rationale of available specificity controls for the *in situ* signal and a discussion of the requirements of the two probe types in their requirements of buffer solutions and various types of pretreatments.

SOURCE OF cDNA PROBES

There are several ways of obtaining nucleic acid probes to be used in *in situ* hybridization. One can clone a cDNA to a particular mRNA at a substantial effort, or obtain clones from others. Such clones can be used either directly (e.g., nick-translated double-stranded DNA) or as a template for producing single-stranded complementary RNA probes. Alternatively, one can synthetically produce a single-stranded, short DNA sequence capable of hybridizing to that mRNA (oligonucleotide probes). Recombinant DNA and microbiological techniques are required to identify and radiolabel the cloned DNAs necessary for *in situ* hybridization experiments. These requirements can often be eliminated by the use of synthetic oligonucleotides complementary to the mRNA of interest. Compared with nick-translated cloned cDNA probes, oligonucleotides are easy to manufacture, penetrate tissue much more readily, can be made to correspond to a DNA sequence at any point in a known structure, and allow for the design of more precise technical controls for *in situ* hybridization studies. RNA probes, on the other hand, are easily produced from DNA clones, can be labeled to high specific activity, are single-stranded, can be reasonably controlled for sequence and length, and can be easily treated with RNase to control for specificity and signal-to-noise ratio.

PROBE COMPOSITION AND HYBRID STABILITY

The stability of any given cDNA–RNA hybrid is largely determined by five variables: (1) probe length, (2) probe composition (%[GC] base pairs versus [AT] or [AU] base pairs), (3) temperature, (4) salt concentration, and (5) number of base-pair mismatches. Hybrid strength is weakened with increasing wash or hybridization temperature, decreasing salt concentration, decreasing duplex length, decreasing percentage of GC pairing (GC pairs have three hydrogen bonds compared with two for AT and AU pairs), and increasing number of mismatched base pairs. For example, an oligonucleotide with a 65% GC content will form a more stable duplex with its complementary strand at any given temperature than an oligomer with an equal length but 50% GC content. One can therefore determine a temperature at which the former oligomer will remain complexed to its mRNA, and the latter oligomer will dissociate (melt). The temperature at which a DNA–RNA hybrid population is half dissociated is defined as the T_m (melting temperature) value and is, to a certain extent, an index of stability. The same logic is true for short versus long probes, with long probes being more stable than short probes.

LABELING OF OLIGONUCLEOTIDE PROBES

Once an oligonucleotide has been synthesized and purified, it can be radiolabeled to permit its detection following hybridization to a complementary nucleic acid sequence (e.g., a specific mRNA). The use of oligonucleotides as hybridization probes for screening clones from cDNA libraries, or Southern (DNA) and Northern (RNA) blots, requires probes of a high specific activity. Because these applications mainly require a high specific activity (for sensitivity) rather than a high degree of resolution (i.e., only that provided by x-ray film), the probes are generally [^{32}P]-labeled on the 5′ hydroxyl terminus using T_4 polynucleotide kinase with [gamma-^{32}P]ATP as the [^{32}P]donor (Maniatis et al., 1982). This procedure provides labeled probes suitable for low-resolution, high-signal screening of libraries or gels. This 5′ hydroxyl terminal reaction can be used only for [^{32}P]- and [^{35}S]-labeling (Beltz and O'Brien, 1981). Because we were interested in using lower-energy beta-emitters (e.g., [³H]) for high-resolution autoradiography, an alternative method of labeling the probes was developed (Lewis et al., 1985). The enzyme terminal deoxynucleotidyl transferase catalyzes the polymerization (multiple addition) of deoxynucleoside triphosphates (dNTPs) to the 3'-hydroxyl terminus of DNA (Bollum, 1974), a reaction used for complementary homopolymer (same nucleotide) tailing of vectors and insert sequences for DNA cloning (Maniatis et al., 1982). With this reaction, we have added homopolymers or random 3'-tails to oligonucleotide probes (Figure 7-3) using [³H], [³⁵S], or [³²P]dNTPs, or biotinylated dUTP, for *in situ* hybridization histochemistry and Northern blotting (Lewis et al., 1984, 1986a, 1986b, in preparation). This approach has the advantages of labeling the oligonucleotide probes with multiple radioactive bases, to enhance detection as well as permit the use of lower energy emitters (e.g., [³H]) to enhance anatomical resolution (Figure 7-4).

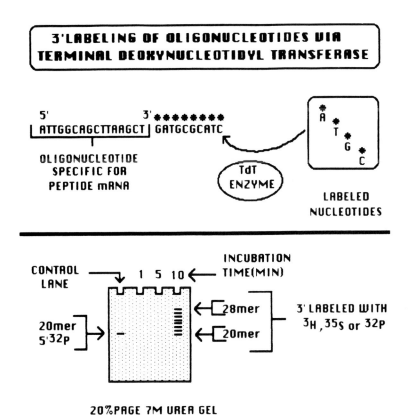

Figure 7-3 Radioactive deoxynucleotides are sequentially attached to the 3' end of a specific oligonucleotide with terminal deoxynucleotidyl transferase (TdT). The reaction mix is incubated at 37° C for a specific time, terminated with EDTA, and run on a 20% PAGE 7M urea gel. The number of nucleotide additions, which increases with time of incubation, can be determined from the gel.

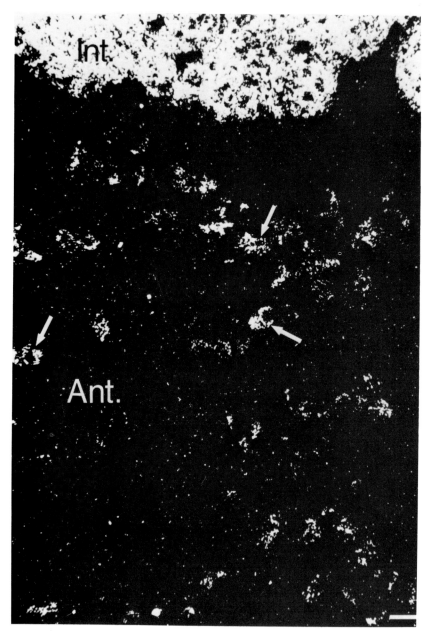

Figure 7-4   Darkfield photomicrograph of rat pituitary hybridized with 3′ [³H]-labeled POMC oligonucleotide. Intermediate lobe (Int.) is heavily labeled while only a subpopulation of cells in the anterior lobe (Ant.) are labeled (*arrows*). Bar = 25 μm.

IN SITU HYBRIDIZATION SPECIFICITY CONTROLS

In immunocytochemistry, the specificity of the antisera used is a major concern because cross-reactivity can result in misleading false-positive staining. In a parallel fashion, hybridization of a probe to noncomplementary RNA (or other forms of nonspecific binding) would yield erroneous localizations in an *in situ* hybridization study. Thus, as described next, several approaches to controls for *in situ* hybridization have been developed to ensure the specificity of the detected hybrid signal.

Combined *In Situ* Hybridization and Immunocytochemistry

The availability of antisera specific to peptides or proteins coded for by a particular mRNA is helpful in establishing specificity of localization. With the proper fixation conditions, both peptide antigenicity and mRNA hybridizability are maintained. Correlative immunocytochemistry and *in situ* hybridization histochemistry have been carried out on serial sections (Gee et al., 1983) and even on the same section (Griffin et al., 1983). If serial sections are used, the section processed for mRNA localization can be taken between two sections processed for peptide immunocytochemistry, using either the same antiserum or antisera directed against different peptides from the same precursor coded for by the mRNA (see Watson and Akil, 1983, for further discussion). When processing a single section for both histochemical procedures, immunocytochemistry may be carried out using the peroxidase–antiperoxidase (PAP) method (Watson and Akil, 1983) in the presence of ribonuclease A inhibitors, followed by *in situ* hybridization and emulsion dipping; autoradiographic grains and horseradish peroxidase reaction product are then detected simultaneously using darkfield or brightfield microscopy (Figure 7-5). Alternatively, the hybrids can be formed first, the immunocytochemistry developed next, and then the dipping of the section in photoemulsion for visualization of the *in situ* signal.

Competition Studies

Competition blocking studies can be carried out by using saturating concentrations of unlabeled cDNA oligonucleotide to hybridize to the specific mRNA in the tissue section before the labeled cDNA probe is applied. Because the concentration of specific mRNA is small relative to potential "nonspecific hybridization" sites, the specific hybridization signal should be the most attenuated. This has been demonstrated in our laboratory for a variety of peptide mRNA probes, such as an alpha-MSH probe hybridizing to pituitary intermediate lobe (Figure 7-6), or an arginine–vasopressin (AVP) probe hybridizing to supraoptic nucleus.

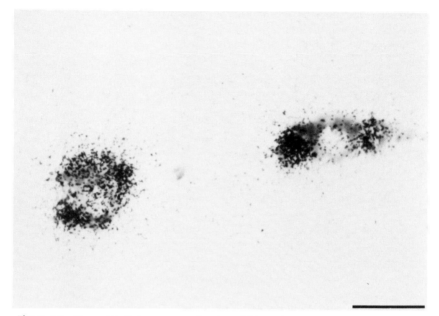

Figure 7-5 Dynorphin B immunocytochemistry (ICC) and AVP *in situ* hybridization in rat accessory nucleus of the same tissue section. ICC was carried out first using the PAP method in the presence of 2 m*M* vanadyl-ribonucleoside complexes (VRC) followed by *in situ* hybridization with [^{35}S]-labeled AVP oligomer. Sections were then emulsion-dipped in NTB-2 (Kodak) and exposed for 11 weeks before development. AVP *in situ* grains are colocalized with Dyn B ICC–stained cells. Grains cluster over the cytoplasm rather than the nucleus. (Bar = 25 μm.)

Multiple Hybridization Probes

Multiple probes complementary to different regions of the same mRNA can be used to hybridize to the same cells, providing additional evidence for the specificity of the hybridization. When applied to the same section, the use of such probes can result in an enhanced cellular signal. Ongoing studies in our laboratory with five different oligonucleotides (each 30 mer) specific for three different regions of the rat prodynorphin mRNA (in collaboration with J. Douglass, O. Civelli, and E. Herbert) indicate that each demonstrates correct anatomical distribution within the hypothalamus. Use of these oligomers in unison should provide the amplification necessary for comparative visualization of this low-abundance mRNA with that of AVP mRNA in same-section or serial section analyses.

Thermal Stability of *In Situ* Hybrids

An important physical characteristic of DNA–RNA (or RNA–RNA) hybrids is their mean thermal denaturation temperature—that is, the T_m, the temperature at which 50% of the hybrids dissociate (see Figure 7-7). Melting temperatures are modifiable, and depend on such factors as salt concentration, percent G-C content, length, and the number of base-pair mismatches (Bonner et al., 1973; Britten et al., 1974; Cantor and Schimmel, 1982). Because base-pair mismatches reduce the thermal stability of hybrids, a melting temperature analysis can be used to determine whether or not a labeled probe is hybridizing to the appropriate complementary sequence *in situ* (Szabo et al., 1977; Kelsey et al., 1986). T_m values obtained from *in situ* studies have been compared with those obtained from solution or solid phase DNA–RNA hybridization studies of extracted RNA (Szabo et al., 1977; Kelsey et al., 1986) and can be compared with calculated theoretical T_m values (Kelsey et al., 1986), using the formula $T_m = 16.6$ log Na/probe length (see Bonner et al., 1973; Britten et al., 1974; Cantor and Schimmel, 1982). For example, proopiomelanocortin mRNA is present in the intermediate but not the posterior lobe of the rat pituitary, so hybrids in the former should have an apparent T_m similar to the theoretical T_m, whereas mismatched "hybrids" in each lobe are thermally unstable and should have a much lower apparent

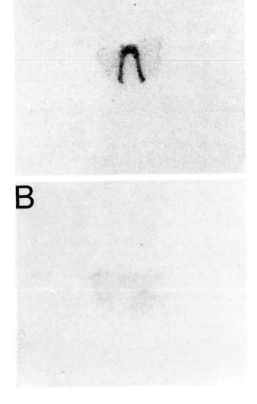

Figure 7-6 (A) X-ray film autoradiograph of rat pituitary section hybridized *in situ* with [³²P]-labeled POMC oligomer. (B) Autoradiograph of rat pituitary section that was initially hybridized with 50 ng of unlabeled POMC oligomer, washed, and then hybridized with 10 ng of [³²P]-labeled POMC oligomer. Intermediate-lobe-specific signal is completely absent whereas the anterior and posterior lobe signals are reduced.

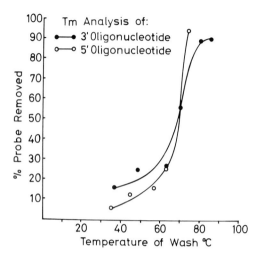

Figure 7-7 A 50% melting temperature (T_m) analysis of 5' and 3' [^{32}P]-labeled rat POMC oligonucleotide. The 3' radioactive tail added to the oligo was between 5 and 15 bases. After *in situ* hybridization and x-ray exposures, tissues were washed in 0.5 × SSC for 15 minutes at temperatures ranging from 37 to 85° C. Pre- and postmelt x-ray films were compared by densitometry to determine the percentage of probe lost at each temperature. The measured T_m values were very similar to the calculated theoretical value of 69° C.

T_m. The true hybrid melting curve has a characteristic sigmoid shape, while the nonspecific hybrid signal decreases linearly with increasing temperature (the latter is illustrated in the posterior pituitary signal). We obtained the results (for both specific and nonspecific hybrids) described earlier using an oligonucleotide complementary to the alpha-MSH coding region of pro-opiomelanocortin (Lewis et al., 1985), confirming the usefulness of T_m analysis in discriminating specific and nonspecific hybridization *in situ*. In other studies, we have found that the addition of a heteropolymer tail to the 3'-end of the oligonucleotide in the terminal deoxynucleotidyl transferase labeling reaction altered the hybridization properties (T_m) of the probe by only a few degrees.

Finally, the T_m value for an oligonucleotide can indicate which temperature maximizes the signal for that specific hybridization, compared with background. An optimal hybridization temperature of T_m −20 to 25° C is commonly suggested as a good starting point. The optimal hybridization temperature of 45° C for our alpha-MSH oligomer stemmed from the experimentally determined T_m of about 71° C (Figure 7-7).

Northern Gel Analysis

Determination of specific mRNA size (with Northern gels; Thomas, 1980) and the abundance in the tissue of interest with the same oligonucleotide probe used in anatomical preparations provide a good index of hybridization specificity. We have been able to show that intact mRNA of the correct size can be isolated from 4% formaldehyde-fixed tissue and quantitated by both dot-blot hybridization and Northern analyses (Kelsey et al., 1984; 1986). Furthermore, quantitative Northern analysis is the method of choice for the confirmation of *in situ* hybridization-detected mRNA level changes, seen in development or with drug and physiologic manipulations (Sherman et al., 1986).

Because of the relatively low specific activities attainable with [^{32}-P]-kinased oligomers compared with either nick-translated DNA clones or M13-primed probes, satisfactory results are difficult to obtain with rare mRNAs; however, quantitative results with moderately abundant RNAs have proven informative. For example, rat vasopressin mRNA is readily detected in hypothalamic RNA using the AVP II 26 base oligonucleotide (Sherman et al., 1986); yet, when using either of three different 30-base oligomers to rat dynorphin mRNA, which is a low-abundance RNA (<1% of VP mRNA) present in the same tissue (and same cells), the results are negative. Both mRNAs, however, are readily detected with nick-translated probes or by *in situ* hybridization (Sherman et al., 1984).

ALTERNATIVE APPROACHES

SP6 cRNA Probes

The first part of this chapter focused on the use of oligonucleotides for *in situ* hybridization. The constant length and content of these monospecific probes provides several major advantages in establishing specificity controls and determining signal-to-noise ratio. If one is prepared to commit greater effort in the techniques of molecular genetics and a modest effort in microbiology, it is possible to produce large amounts of complementary RNA probes of even higher specific activity, and constant length (Green et al., 1983; Cox et al., 1984; Johnson and Johnson, 1984; Shaw et al., 1985). Figure 7-8 shows the basic structure of the commercially available SP6 Riboprobe system. In brief, a double-stranded DNA clone, coding for the specific mRNA, is inserted in the SP6 vector. The SP6 vector has a unique bacterial RNA polymerase binding site. When the correct bacterial RNA polymerase is used, the DNA is transcribed into single-stranded RNA. Because an *in vitro* reaction occurs, labeled ribonucleotides can be incorporated efficiently into this RNA transcript. Two other major advantages are found in the SP6 Riboprobe system. First, the RNA polymerase copies the DNA 10 to 15 times per reaction, resulting in large quantities of labeled RNA probes. The second advantage is that the double-stranded DNA template can be cut (by restriction endonucleases) to produce a specific length RNA transcript. On the other hand, from a technical point of view, there are several problems with the system (or at least major considerations). First, since the production of such probes requires subcloning of existing cDNAs, general molecular biology and microbiological skills are needed. Second, single-stranded RNA probes are vulnerable to the degradative ribonucleases. Third, RNA probes form more stable hybrids with mRNA and require higher hy-

RNA PROBE PRODUCTION:SP-6 SYSTEM

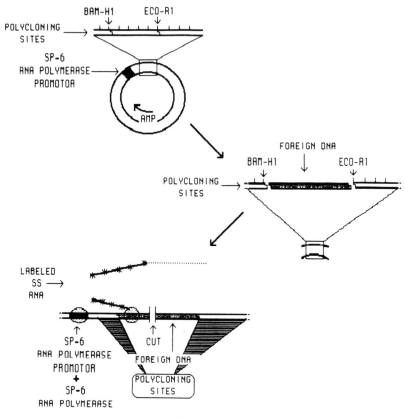

Figure 7-8

bridization temperatures (15–30° C more, which may damage tissue sections) to achieve stringent conditions. Finally, longer RNA probes may be attractive for some purposes (e.g., Northern analysis) but may have to be partially hydrolyzed for *in situ* hybridization (Cox et al., 1984). The increase in specific activity and probe length results in a much more highly labeled, and therefore more useful, probe. Figure 7-9 shows an example of proenkephalin mRNA localization at the level of the basal ganglia (including the parvocellular elements of the caudate).

Controls for RNA Probes

RNase Treatment

After the labeled cRNA probe is applied to the tissue, the tissue can be exposed to RNase A (10 µg/50× in 0.5 M NaCl, 10 mM Tris, pH 8, at 37° C for 20 minutes). This enzyme is specific for single-stranded RNA, and is thus able to hydrolyze free cRNA and mRNA in the tissue. The main result is the dramatic reduction of nonspecific (background) signal; leaving the double-stranded (specific) RNA–RNA hybrid in the tissue.

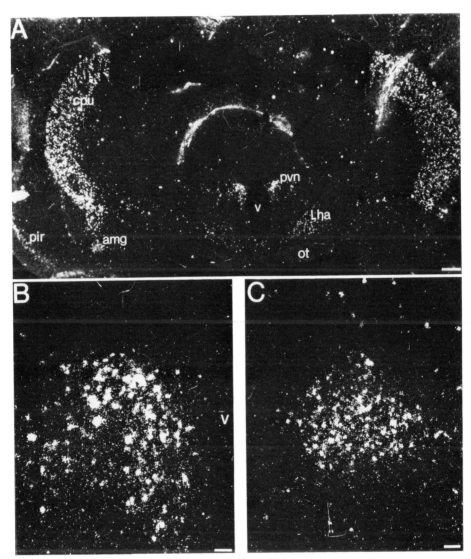

Figure 7-9 (A) Darkfield photomicrograph of rat brain hybridized with [^{35}S]-labeled SP6 en-kaphalin riboprobe (900 b). Concentrations of grains are localized over the caudate putamen (cpu), piriform cortex (pir), amygdala (amg), paraventricular nucleus (pvn), and lateral hy-pothalamic area (Lha). The asterisks (*) represent artifacts resulting from tissue folds. (ot = optic tract; v = third ventricle, bar = 500 μm.) (B) High-power magnification of pvn. (V = ventricle, bar = 50 μm.) (C) High-power magnification of amygdala. (Bar = 50 μm.)

Messenger RNA Sense Riboprobe

Using the Riboprobe system, it is possible to produce a labeled single-stranded RNA probe with the same sense as the natural mRNA (as opposed to an antisense strand, e.g., cRNA). Such a probe should not hybridize to endogenous mRNA. After RNase treatment, tissue previously treated with this control probe should give the most accurate estimate of nonspecific hybridization.

Blocking by Unlabeled cRNA

By preparing unlabeled specific cRNA, and preincubating tissue with it, it is possible to saturate that specific complementary area of the endogenous mRNA. With such pretreated tissue, the labeled cRNA probe should be unable to form a hybrid.

Comparison of Hybridization Conditions for Oligonucleotide and RNA

Probes. The two main types of probes discussed in this chapter are the short single-stranded cDNA oligonucleotide probes, and the potentially much longer single-stranded cRNA probes. Their chemical differences, apparent differences in "stickiness," length variation, and differences in conditions necessary to produce a hybrid all argue for the importance of adjusting hybridization conditions to the type of probe used. The following section details our current method for optimizing the oligonucleotides. We are currently investigating the changes necessary to optimize for long cRNA probes. One clear area of difference is that of tissue penetration. Oligonucleotides do not seem to require permeabilization of the tissue (with HCl or proteinase K), whereas such a step is critical for the longer RNA probes. Other areas for active investigation include random addition of nucleic acid, prehybridization buffers, and exposure to cold ethanol. More important than this specific method is the use of continuing exploration of the methods and logic of *in situ* hybridization.

IN SITU HYBRIDIZATION PROTOCOL

Tissue Preparation

1. Perfuse the animal with 4% formaldehyde (from paraformaldehyde) in 0.1 M NaPO$_4$ buffer (approximately 1 liter) for 30 minutes. Remove the desired tissue and postfix for 1 hour in formaldehyde solution. Place it in 15% sucrose (in phosphate-buffered saline) overnight. Freeze the tissue in liquid nitrogen and store at $-80°$ C.
2. Section the tissue at $-20°$ C (10 μm or less in thickness) and place the sections on gelatin-subbed slides. Store the tissue sections at $-80°$ C.

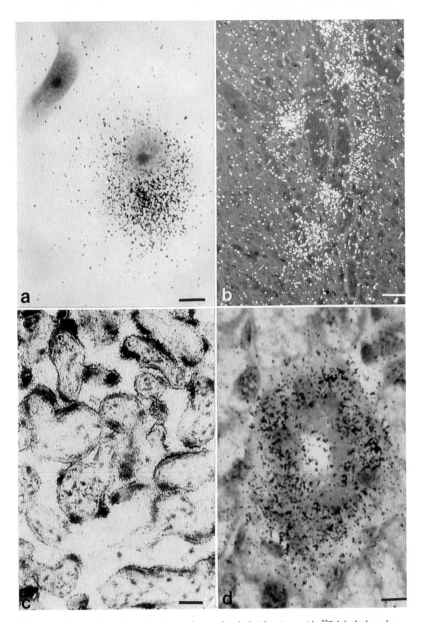

Plate 1. Autoradiographs with K5 emulsion after hybridization with ^{32}P-labeled probes. (a) Vero cells from a culture infected with Ross river virus (RRV), hybridized with a 30-mer oligodeoxyribonucleotide probe prepared to the E2 region of RRV 26S RNA. (Bar = 7.0 μm.) (b) Photomicrograph in darkfield of the left ventricle in a 5-μm frozen section of rat heart hybridized with a 30-mer oligodeoxyribonucleotide probe corresponding to the cardiodilatin coding region of mRNA for rat atrial natriuretic peptide (rANP). Labeled cells are myocytes surrounding an arteriole. (Bar = 40 μm). (c) A 5-μm frozen section of human placenta at term after hybridization with a cDNA probe 621 bp for the α subunit of human chorionic gonadotrophin (hCG). Labeled areas show the location of mRNA for CG in syncytiotrophoblasts of fetal cotyledons. (Bar = 80 μm). (d) A 3-μm paraffin section of submandibular gland from a female Swiss mouse. mRNA for mouse glandular kallikrein (mGK) has been located to cells of the granular convoluted tubule using a single-stranded RNA probe prepared in the SP-6 vector system from the 498-bp cDNA pMK-1. (Bar = 10.0 μm.)

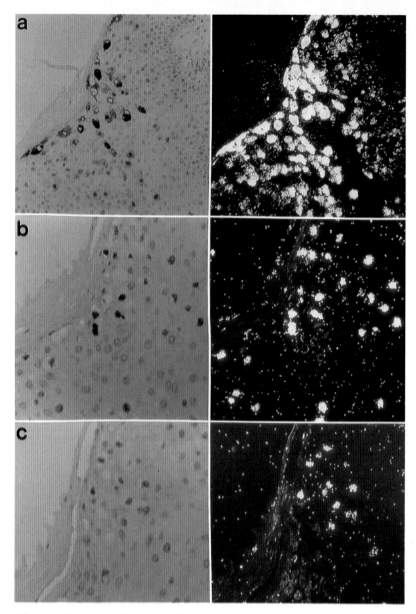

Plate 2. Detection of human papilloma virus (HPV) DNA and RNA in condyloma acuminatum. Sections of condyloma acuminatum (genital wart) were fixed in formalin and processed for paraffin sectioning by routine automated procedures. *In situ* hybridization was carried out using ^{3}H-labeled RNA probes corresponding to HPV11 sequence, of specific activity 1.1×10^8 dpm/μg and at probe concentrations calculated to give signals 10-fold lower than at saturation. Exposure time was 4 weeks. Three adjacent sections are shown. Magnification is 132×. The right panels are darkfield images of the same sections shown in brightfield illumination in the left panels: (a) Detection of mRNAs complementary to an HPV11 antisense sequence probe. The DNA in the tissue was not denatured. Note that the signal is much higher over cytoplasm than over nuclei. (b) Detection of HPV DNA. The DNA in the tissue was denatured, as described in the text, and sections were hybridized with a sense-strand probe. Hybridization is confined to nuclei. (c) Hybridization of the sense-strand probe to tissue in which DNA had not been intentionally denatured. Hybridization is markedly lower than after denaturation of DNA, being confined to the more heavily labeled cells shown in (b). The low signals observed may result from denaturation of DNA by formaldehyde during the fixation procedure (see text).

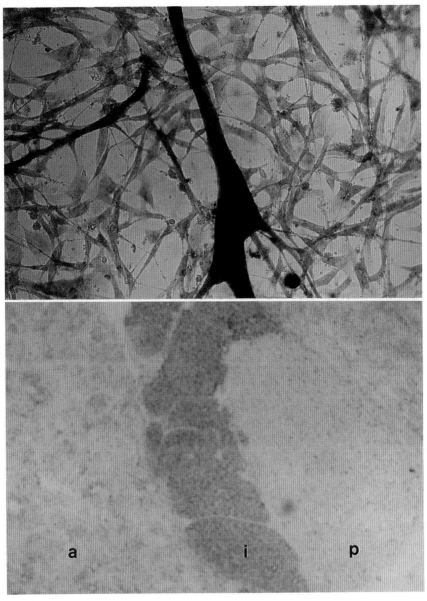

a   i   p

Plate 3. (a) Histochemical detection of muscle-specific gene expression in cell cultures. A recombinant probe containing the 3' untranslated region of alpha cardiac actin (300 nucleotides, see Paterson and Eldridge, 1984) was biotinylated and hybridized, as described, to differentiating muscle cultures. The biotinated DNA was detected by alkaline phosphatase, as shown in Figure 4-7. The low-power photomicrograph (160× at the 35-mm negative) clearly indicates only the myotubes in the culture. (b) Histochemical detection of proopiomelanocortin mRNA in rat pituitary. Pituitaries were isolated from adult male rats and frozen in isopentane–liquid nitrogen. They were sectioned at 10-μm and hybridized to a biotinylated POMC probe (supplied by Julie Jonassen; see Roberts et al., 1982). The probe was detected by the alkaline procedure (see Figure 4-7). The enlargement on the photograph is 1100×. The anterior pituitary (a) shows some positive cells, the intermediate (i) is heavily labeled, with all cells positive, and the posterior (p) is negative.

a

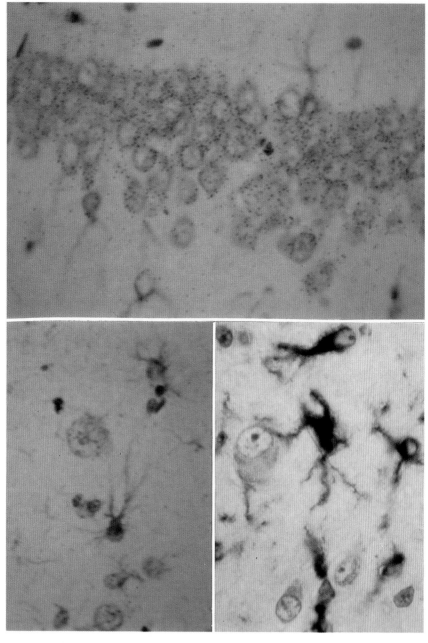

b

c

Plate 4. (a) Hippocampus from adult rat (Bouin's, fixed for 48 hours) *in situ* hybridized with [³H]-poly-U, as in Figure 5-1, and immunoreacted with antibody to GFAP, as in Table 5-3. (b, c) A comparison of immunohistochemical localization of glial fibrillary acidic protein immunoreactive product in rat cerebellum stained with Meyer's hematoxylin (b) or Lapham's stain (c).

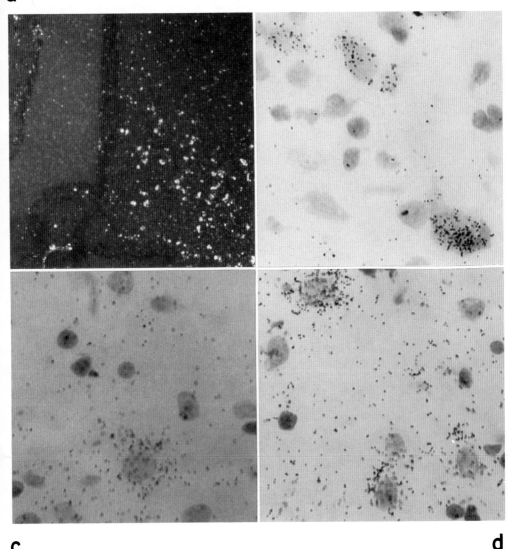

a

b

c

d

Plate 5. (a) Positive chemography in a paraformaldehyde-fixed section of rat hypothalamus. An adult rat was perfusion-fixed with 3% paraformaldehyde in PBS. Following decapitation of the animal, the hypothalamus was dissected out of the brain and immersion-fixed in the paraformaldehyde solution for an additional hour, then equilibrated in a 30% (w/v) solution of sucrose in PBS at 4° C. The tissue was frozen and cut into sections of 8 μm. The sections were incubated in hybridization buffer without radioactive probe overnight at room temperature, and washed as described in Appendix A. The slides were coated with NTB2 emulsion (diluted 1:1 with 600 mM ammonium acetate, pH 7.0), and stored at 4° C for 1 month, then photodeveloped and stained with cresyl violet. The darkfield photomicrograph of a coronal section through the hypothalamus shows extensive silver grain reduction over cells in the mediobasal portion of the tissue. Frequently, clusters of reduced silver grains were also seen in the more dorsal region of the hypothalamus, over cells subadjacent to the dorsal part of the third ventricle (not shown). (b, c, and d) A comparison of signal strength and signal-to-noise ratio in hypothalamic sections probed with POMC probes labeled with ^{3}H, ^{32}P, and [35]S. Rat hypothalamic sections were hybridized with either ^{3}H-cDNA (Plate 5b; 1.25 ng, s.a. 6 × 10[6] cpm/μg, 51 days of exposure), ^{32}P-cRNA (Plate 5c; 0.09 ng, s.a. 4.4 × 10[8] cpm/μg, 8 days of exposure), or ^{35}S-cRNA (Plate 5d; 0.2 ng, s.a. 10[8] cpm/μg, 14 days of exposure).

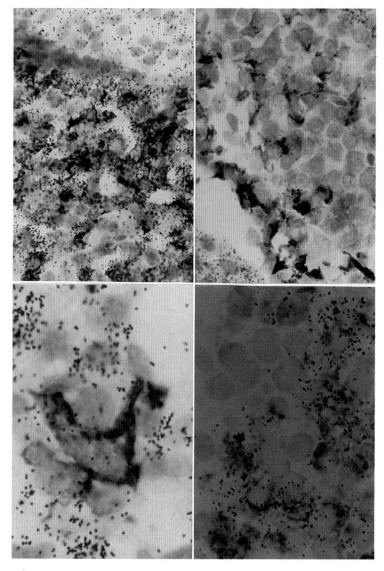

a

b

c

d

Plate 6. Detection of peptide hormones and peptide hormone mRNAs on the same rat pituitary tissue section. Sections of paraformaldehyde perfusion-fixed rat pituitaries were subjected to immunocytochemical treatment for antigen detection, then to *in situ* hybridization for mRNA detection. Solutions of primary and secondary antisera contained 0.04% DEP. Antibody localization used the avidin-biotin-horseradish peroxidase system with diaminobenzidene as the indicator chromagen (seen as the brown color). mRNAs were localized with ³H-cDNA probes. Further experimental details are described elsewhere (Shivers et al., 1986). (a) Antirat PRL antiserum and PRL cDNA detect the corresponding protein and mRNA within the same cells in the anterior pituitary. No labeling with either probe is seen in the intermediate or posterior lobes of the tissue (not shown). (b) Anti-PRL antiserum detects product in a large number of anterior pituitary cells. POMC cDNA probe reacts with a different subset of anterior pituitary cells and with all cells of the intermediate lobe. (c) Anti-ACTH antiserum detects product in a small subset of anterior pituitary cells. PRL cDNA probe reacts with numerous anterior pituitary cells, but these two probes show no overlap. (d) Colocalization of PRL and PRL mRNA in a fresh-frozen, postfixed section of rat pituitary. Tissue was prepared as described in Appendix A, incubated with anti-PRL antiserum (1:1000 dilution) in the presence of DEP.

a

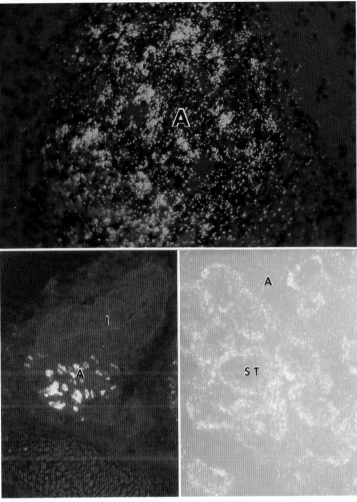

b   **c**

Plate 7. (a) Hybridization of embyronic e17 rat pituitary using ³H-POMC cRNA.
Cryostat sections from perfusion-fixed rat e17 embryo were hybridized with
³H-POMC cRNA prepared from a vector containing POMC-118 DNA. Sections
were treated with 1 μg/ml of proteinase K for 10' at room temperature, washed,
treated with triethanolamine/acetic anhydride (Cox et al., 1984), washed, and
dehydrated. The sections were prehybridized and hybridized with buffers as in
Figure 10-2. They were hybridized at 45° C overnight in 50% formamide and
washed to a stringency of 0.2 × SSC at room temperature. The sections were then
exposed for 1 week, after coating with NTB-2 emulsion. Heavily labeled, single
cells are present in the anterior lobe (A) (the intermediate lobe is not present in
the section). (b) Immunocytochemical detection of POMC-peptides at e14. Immunopositive cells are located only in the ventral part of the anterior lobe (A),
similar to the distribution of POMC mRNA containing cells seen in Plate 7a
(I = intermediate lobe). (c) Hybridization of male submaxillary glands with
³⁵S-NGF DNA. NGF cRNA was radiolabeled with ³⁵S-nucleotides and hybridized to sections containing male mouse submaxillary gland. Following hybridization and washing, the sections were coated with emulsion (NTB-2) and exposed for 2 days at 4° C. Labeling is seen over the secretory tubules (ST), but
is absent from acini (A).

Plate 6 (continued).   ³⁵S-PRL cRNA was used as described in Appendix A.
Note that, although tissue morphology is somewhat poorer than in perfusion-
fixed tissue such as that seen in (a) through (c), both the immunocytochemistry
and *in situ* hybridization reactions give the expected distribution of signals.

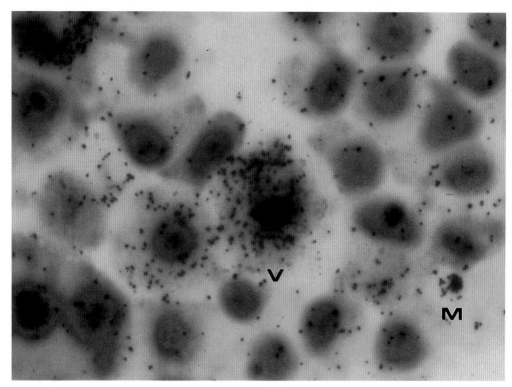

Plate 8. Vero monkey kidney cells and sheep choroid plexus cells infected with measles (M) and visna (V) viruses, respectively. Measles virus is detected with ³H, and the grains appear magenta, whereas visna is detected with ³⁵S, producing cyan grains in the second layer of the emulsion.

Prehybridization Conditions

1. Warm the tissue sections to room temperature for at least 30 minutes (to allow the sections to adhere to the slide), and place a small dot of fingernail polish near the tissue and allow to dry thoroughly. The fingernail polish will support the coverslip used in the final hybridization steps.
2. Deproteinate sections in 50 μl of proteinase K (1 μg/ml) for 30 minutes at 37° C. (Alternatively, use 0.2 N HCl for 10 minutes at room temperature.)
3. Wash section for 10 minutes in 2 × SSC (300 mM NaCl and 30 mM sodium citrate) at room temperature.
4. Remove excess 2 × SSC and apply 50 μl of hybridization buffer to each section for at least 1 hour at the appropriate hybridization temperature (varying with the probe type used). The hybridization buffer contains 50% formamide, 10% dextran sulfate, 3 × SSC, 1 × Denhardt's (0.02% ficoll, 0.02% polyvinyl pyrrolidone, 10 mg/ml BSA), 100 μg/ml of yeast tRNA, and 100 μg/ml of sonicated salmon sperm DNA.

Hybridization

1. Dilute radioactive DNA probes with a hybridization buffer to give the desired number of cpms per 50 μl. The number of cpms used varies between 1000 and 5 million, depending on probe efficiency and the actual amount of probe (in nanograms) to be applied.
2. Denature double-stranded probes along with salmon sperm DNA by heating at 90 to 100° C for 5 to 10 minutes, then cooling rapidly on ice. The probe is then diluted in the hybridization buffer.
3. Drain nonradioactive hybridization buffer from each section and apply 50 μl of the probe/hybridization mix. Coverslip the sections with a corner of coverslip resting on fingernail polish. Place the slides on wet foam in sealed boxes.
4. Incubate the sections at the desired temperature for 24 hours. The temperature used will depend on the probe length and 50% melting temperature (T_m) of the probe. Initial studies should be done at room temperature.

Posthybridization

1. Carefully remove coverslips by dipping them in 2 × SSC and wash the sections for 30 minutes in 2 × SSC. Drain and wash the slides in 1 × SSC for 30 minutes. Again drain and wash the slides in 0.5 × SSC for 15 minutes at room temperature, 0.5 × SSC for 1 hour at the

hybridization temperature, and 0.5 × SSC for 30 minutes at room temperature.

2. Drain the slides and allow the sections to dry thoroughly at room temperature.

Detection

1. Place the sections to which a [^{32}P] or [^{35}S] probe have been applied on Kodak XAR-5 x-ray film and exposed them for 4 to 48 hours.
2. Dip the sections to which a [^{3}H] probe have been applied in Kodak NTB-2 emulsion (diluted 1:1 with distilled water) and store them desiccated in light-tight boxes at 4° C. Develop test slides at intervals and examine them under darkfield illumination. [^{32}P], [^{35}S], and [^{125}I] probes may also be detected with this emulsion.

FUTURE DIRECTIONS

Much work remains to make *in situ* hybridization histochemistry as easy and logical as immunocytochemistry. Such efforts are actively underway and are beginning to yield a powerful method for all of biology, including molecular neuroscience. Finally, the real need is for accurate single-cell quantitation of mRNA in an anatomic context; it is in this area that progess is most needed.

ACKNOWLEDGMENTS

This paper was supported by NIMH grant MH39717 (SW), MH09059 (JK) NIADDK grant AM34933 (SW), NIDA grant DA02265 (HA), the Theophile Raphael Fund to SJW and HA, and NIMH postdoctoral fellowship MH09239 to TGS.

REFERENCES

Beltz, W.R. and O'Brien, K.J. End labeling of DNA restriction fragments using [S-35]Adenosine 5'-(Gamma-Thio)Triphosphate followed by mercury affinity chromatography. *Fed Proc.* 40: 1849, 1981.

Bollum, F.J. Terminal deoxynucleotidyl transferase. In *The Enzymes*, Vol. 10. P.D. Boyer (Ed.). New York: Academic Press, 1974.

Bonner, T.I., Brenner, D.J., Neufeld, B.R., and Britten, R.J. Reduction in the rat of DNA reassociation by sequence divergence. *J. Mol. Biol.* 81: 123–135, 1973.

Brahic, M. and Haase, A.T. Detection of viral sequences of low reiteration frequency by *in situ* hybridization. *Proc. Natl. Acad. Sci. USA* 75: 6125–6129, 1978.

Britten, R.J., Graham, D.E., and Neufeld, B.R. DNA sequence analysis by reassociation. In *Methods in Enzymology,* Vol. 29. L. Grossman and K. Moldave (Eds.) New York: Academic Press, 1974, pp. 363–418.

Cantor, C.R. and Schimmel, P.R. *Biophysical Chemistry.* San Francisco: W.H. Freeman, 1982.

Cox, K., Deleon, D., Angerer, L., and Angerer, R. Detection of mRNAs in sea urchin embryos in *in situ* hybridization using asymmetric RNA probes. *Dev. Biol.* 101: 485–502, 1984.

Gee, C.E., Chen, C.L., Roberts, J.L., Thompson, R., and Watson, S.J. Identification of proopiomelanocortin neurons in rat hypothalamus by *in situ* cDNA-mRNA hybridization. *Nature* 306: 374–376, 1983.

Gee, C.E. and Roberts, J.L. Laboratory methods for *in situ* hybridization histochemistry: A technique for the study of gene expression in single cells. *DNA* 2:155–161, 1983.

Green, M., Maniatis, T., and Melton, D. Human beta-globin pre-mRNA synthesized *in vitro* is accurately spliced in xenepus oocyte nuclei. *Cell* 32: 681–694, 1983.

Griffin, W.S.T., Alejos, M., Nilaver, G., and Morrison, M.R. Brain protein and messenger RNA identification in the same cell. *Brain Res. Bull.* 10: 597–601, 1983.

Haase, A.T., Ventura, P., Gibbs, C.J., and Tourtellotte, L. Measles virus nucleotide sequences: Detection by hybridization *in situ*. *Science* 212: 672–675, 1981.

Haase, A.T., Walker, D., Stowring, L., Ventura, P., Geballe, A., Blum, H., Brahic, M., Goldberg, R., and O'Brien, K. Detection of two viral genomes in single cells by double-label hybridization *in situ* and color microradioautography. *Science* 227: 189–192, 1985.

Hafen, E., Levine, M., Garber, R., and Gehring, W. An improved *in situ* hybridization method for the detection of cellular RNAs in Drosophila tissue sections and its application for localizing transcripts of the homeotic gene Antennapedia gene complex. *EMBO J.* 2: 617–623, 1983.

John, H.A., Patrinau-Georgoulas, M., and Jones, K.W. Detection of myosin heavy chain mRNA during myogenesis in tissue culture by *in vitro* and *in situ* hybridization. *Cell* 12: 501–508, 1977.

Johnson, M.T. and Johnson, B.A. Efficient synthesis of high specific activity [^{35}S]-labeled human beta-globin pre-mRNA. *Bio. Techniques* 2: 156–162, 1984.

Kelsey, J.E., Watson, S.J., and Akil, H. Changes in pituitary mRNA levels. *Soc. Neurosci. Abstr.* 10: 359, 1984.

Kelsey, J.E., Watson, S.J., Burke, S., Akil, H., and Roberts, J.L. Characterization of proopiomelanocortin mRNA detected by *in situ* hybridization. *J. Neurosci.,* 6: 38–42, 1986.

Kornberg, T., Sien, I., O'Farrel, P., and Simon, M. The *engrailed* locus of drosophila: *In situ* localization of transcripts reveals compartment-specific expression. *Cell* 40: 45–53, 1985.

Langer-Safer, P.R., Levine, M., and Ward, D.C. Immunological method for mapping genes or Drosophila polytene chromosomes. *Proc. Natl. Acad. Sci. USA* 79: 4381–4385, 1982.

Lewis, M.E., Burke, S., Sherman, T.G., Arentzen, R., and Watson, S.J. *In situ* hybridization using a 3' terminal transferase-labeled synthetic oligonucleotide probe complementary to the alpha-MSH coding region of proopiomelanocortin mRNA. *Soc. Neurosci. Abstr.* 10: 358, 1984.

Lewis, M.E., Khachaturian, H., Schafer, M.K.H., and Watson, S.J. Anatomical approaches to the study of neuropeptides and related mRNA in CNS. In *Neuropeptides in Neurological Disease,* ARNMD, Vol. 64, J.B. Martin (Ed.). New York: Raven Press, 1986a.

Lewis, M.E., Sherman, T.G., Burke, S., Akil, H., Davis, L.G., Arentzen, R., and Watson, S.J. Detection of proopiomelanocortin mRNA by *in situ* hybridization with an oligodeoxynucleotide probe. *Proc. Natl. Acad. Sci. USA* 83, 5419–5423, 1986b.

Lewis, M.E., Sherman, T.G., and Watson, S.J. *In situ* hybridization histochemistry with synthetic oligonucleotides: Strategies and methods. *Peptides* 6 (Suppl. 2): 75–89, 1985.

Lynn, D.A., Angerer, L.M., Brushkin, A.M., Klein, W.H., and Angerer, R.C. Localization of a family of mRNAs in a single cell type and its precursors in sea urchin embryos. *Proc. Natl. Acad. Sci. USA* 80: 2656–2660, 1983.

Maniatis, T., Fritsch, E.F., and Sambrook, J. *Molecular Cloning.* New York: Cold Spring Harbor Laboratory, 1982.

McDougall, J.K., Crum, C.P., Fenoglio, C.M., Goldstein, L.C., and Galloway, D.A. Herpes virus-specific RNA and protein in carcinoma of the uterine cervix. *Proc. Natl. Acad. Sci. USA* 79: 3853–3857, 1982.

McWilliams, D. and Boime, I. Cytological localization of placental lactogen messenger ribonucleic acid in syncytiotrophoblast layers of human placenta. *Endocrinology* 107: 761–765, 1980.

Pfeifer-Ohlsson, S., Goustin, A.S., Rydnert, J., Wahlstrom, T., Byersing, L., Stehelin, D., and Ohlsson, R. Spatial and temporal pattern of cellular *myc* oncogene expression in developing human placenta: Implications for embryonic cell proliferation. *Cell* 38: 585–596, 1984.

Pintar, J.E., Schacter, B.S., Herman, A.B., Dungerian, S., and Krieger, D.T. Characterization and localization of proopiomelanocortin messenger RNA in the adult rat testis. *Science* 225: 632–634, 1984.

Pochet, R., Brocas, H., Vassart, G., Toubeau, G., Seo, H., Refetoff, S., Dumont, J.E., and Pasteels, J.L. Radioautographic localization of prolactin messenger RNA on histological sections by *in situ* hybridization. *Brain Res.* 211: 433–438, 1981.

Royston, M.E. and Augenlicht, L.H. Biotinated probe containing a long-terminal repeat hybridized to a mouse colon tumor and normal tissue. *Science* 222: 1339–1341, 1983.

Saber, M.A., Zern, M.A., and Shafritz, D.A. Use of *in situ* hybridization to identify collagen and albumin mRNAs in isolated mouse hepatocytes. *Proc. Natl. Acad. Sci. USA* 80: 4017–4020, 1983.

Scheller, R., Kaldany, R., Kreiner, A., Mahon, C., Nambu, R., Schaefer, M., and Taussig, R. Neuropeptides: Mediators of behavior in Aplysia. *Science* 225: 1300–1308, 1984.

Shaw, P., Sordat, B., and Schibler, O. The two promoters of the mouse alpha-amylase gene Amy-1 are differentially activated during parotid gland differentiation. *Cell* 40: 907–912, 1985.

Sherman, T.G., Akil, H., and Waton, S.J. Vasopressin mRNA expression: A northern and *in situ* hybridization analysis. In *Vasopressin*. R. Schrier (Ed.). New York: Raven Press, 1985, pp. 475–483.

Sherman, T.G., McKelvey, J., and Watson, S.J. Vasopressin mRNA. Regulation in individual hypothalamic nuclei: northern and *in situ* hybridization analysis. *J. Neurosci.* 6: 1685–1694, 1986.

Sherman, T.G., Watson, S.J., Herbert, E., and Akil, H. The co-expression of dynorphin and vasopressin: An *in situ* hybridization and dot-blot analysis. *Soc. Neurosci. Abstr.* 10: 358, 1984.

Singer, R.H. and Ward, D.C. Actin gene expression visualized in chicken muscle tissue culture by using *in situ* hybridization with a biotinated nucleotide analog. *Proc. Natl. Acad. Sci. USA* 79: 7331–7335, 1982.

Southern, P.J., Blount, P., and Oldstone, M.B.A. Analysis of persistent virus infections by *in situ* hybridization to whole body sections. *Nature* 312: 555–558, 1984.

Szabo, P., Elder, R., Steffensen, D.M., and Uhlenbeck, O.C. Quantitative *in situ* hybridization of ribosomal RNA species to polytene chromosomes of Drosophila melanogaster. *J. Mol. Biol.* 115: 539–563, 1977.

Thomas, P.S. Hybridization of denatured RNA and small DNA fragments transferred to nitrocellulose. *Proc. Natl. Acad. Sci. USA* 77: 5201–5205, 1980.

Watson, S.J. and Akil, H. Immunocytochemistry of peptides. In *Current Methods in Cellular Neurobiology, Volume 1: Anatomical Techniques*. J.L. Barker and J.E. McKelvy (Eds.). New York: John Wiley and Sons, 1983, pp. 111–113.

Wolfson, B., Manning, R.W., Davis, L.G., Arentzen, R., and Baldino, F. Co-localization of corticotropin releasing factor and vasopressin mRNA in neurons after adrenalectomy. *Nature* 315: 59–61, 1985.

8

In Situ Hybridization for Mapping the Neuroanatomical Distribution of Novel Brain mRNAs

Gerald A. Higgins and Michael C. Wilson

The mammalian central nervous system (CNS) contains many thousands of different neuronal cell types, which, through a combination of anatomical location and connections, are organized into systems that mediate the various behavioral states of the organism. The differences between cell types, and the establishment and maintenance of synaptic connections among functionally related neuronal subpopulations, must arise from the differential expression of sets of genes.

Classically, neurons have been differentiated on the basis of morphological criteria, including cell shape, size, location, staining characteristics, and more recently, afferent and efferent connections determined using tract-tracing methods. Similarly, immunocytochemical studies have provided evidence for the biochemical heterogeneity of neuronal populations through the localization of defined peptides, enzymes, receptors, and other neuroactive substances, and monoclonal antibodies have been raised that recognize a variety of antigenic determinants present on different neuronal and glial subtypes in the developing and adult CNS (Emson, 1983; Hockfield and McKay, 1985; Levitt, 1984; Sternberger et al., 1982).

The discovery of a large number of heterogeneously distributed brain peptides, coupled with the vast protein-encoding potential of brain (Chikaraishi, 1979; Bantle and Hahn, 1976; Chaudhari and Hahn, 1983), suggests that many cell-type-specific proteins may exist. Preliminary studies of mouse and rat brain RNA sequence complexity estimate that it contains sufficient information to encode 50,000 to 200,000 different polypeptides (Chikaraishi, 1979; Milner and Sutcliffe, 1983). More than 50% of the sequence complexity of brain RNA is represented by rarer mRNAs (Milner and Sutcliffe, 1983), which although less abundant within the whole brain, may be very abundant within restricted cells or populations of cells because of their limited neuroanatomical distribution.

A powerful approach for the identification and characterization of cell-type-specific proteins is through the use of cDNA cloning and hybridization methods. cDNA libraries can be constructed that represent all the mRNA sequences present within a given tissue or region. The primary sequences of proteins can be deduced from the nucleotide sequence of the cloned cDNA, and their anatomical distribution mapped with antibodies raised against synthetic peptides predicted from the sequence (Sutcliffe et al., 1983; Bloom et al., 1985). However, this kind of approach requires sequencing of full-length cDNAs for identification of open reading frames, chemical synthesis of peptides, dependence on antigenicity of synthetic peptides for the immune response, and the careful control of antibody specificity necessary for immunocytochemical localization of novel proteins.

A more direct approach for the identification of central neurons expressing a particular novel brain mRNA is through the use of *in situ* hybridization. In the strategy presented here, we have made cDNA libraries from whole mouse brain poly-A RNA, or from regional, substracted RNA populations, and used individual cDNAs for *in situ* hybridization screening of the neuroanatomical distribution of homologous mRNAs. Thus, a given novel brain mRNA can be selected based on its anatomic localization in brain before more complete characterization occurs. Using this approach, we have mapped the neuroanatomical distribution of a brain-specific mRNA, MuBr 8, and show that although it is widely distributed in the CNS, its expression is limited to defined neuronal subpopulations. Additionally, we have used the subtractive hybridization to isolate more regionally circumscribed transcripts, including H-CbE3 and H-CxC4 mRNAs, which are abundant in the hippocampal formation and amygdaloid complex, but not other brain regions.

METHODS

In Situ Hybridization

Sources of Probes

Recombinant cDNA were generated from whole mouse brain poly-A RNA, as described previously (Branks and Wilson, 1986). Individual "brain-specific" cDNAs were used for RNA blot hybridization to assay for their size, abundance, and tissue and regional brain distribution of homologous mRNAs. Two of these cDNAs, which we have used for *in situ* hybridization, will be discussed in this paper: (1) p2, which hybridizes to the 3' portion of alternatively spliced 3.2- and 2.5-kb mRNAs, which encode proteolipoprotein, a marker for oligodendrocytes, and (2) p8, which recognizes a 2-kb mRNA expressed by a number of neuronal cell groups in brain (Branks and Wilson, 1986).

Additionally, a cDNA complementary to vasopressin/neurophysin precursor mRNA, isolated by Schmale and Richter (1984), was used to compare perfused versus fresh tissue hybridization (see Figure 8-2; Chapter 9).

Preparation of Probes for In Situ Hybridization

Two methods were used to prepare probes for *in situ* hybridization. Nick translation of isolated double-stranded cDNA inserts to high specific activity ($2–8 \times 10^8$ cpm/μg for ^{32}P; 1×10^9 cpm/μg for ^{35}S nucleotide precursors), provided convenient probes for *in situ* analysis. Alternatively, cDNA was inserted into plasmid vectors containing RNA polymerase promoters. These vectors contain SP6, opposing SP6 and T7, or T3 and T7 promoters for initiation of synthesis of single-stranded RNA transcripts using the appropriate prokaryotic RNA polymerase. RNA probes offer several advantages for *in situ* localization of mRNA species in tissue sections (see also Chapter 3): (1) RNA–RNA hybrids are thermodynamically more stable than DNA–RNA hybrids; (2) single-stranded probes have no competing complementary strand in solution during hybridization, as double-stranded (i.e., symmetric), nick-translated probes do, for example; (3) signal-to-noise ratio can be enhanced by posthybridization pancreatic RNase digestion of nonhybrid-bound probe; and (4) probe size can be adjusted by selective alkaline hydrolysis.

Isolated cDNA inserts were ligated either into the polylinker sequence of pSP65 (i.e., vasopressin cDNA) or pGEM2 (MuBr 8 cDNA) plasmid vectors (Promega Biotech; Madison, WI). In cases in which the coding strand of the cDNA insert is not known, dual promoter plasmid vectors such as pGEMINI (Promega Biotech; Madison, WI) allow the determination by hybridization of the polarity of the mRNA. They can also be used to generate sense-strand RNA for control of the specificity of *in situ* hybridization (see Figure 8-3).

Transcription of RNA Probes

Plasmids were linearized by cleavage with appropriate restriction enzymes within polylinker sequences but distal to the site of RNA synthesis. We did not use enzymes that produce 3′ overhangs, as this may cause aberrant transcription of large-molecular-weight transcripts. Occasionally we had difficulty regenerating restriction sites adjacent to the site into which the insert was ligated; in these cases other unique sites downstream of the polylinker, but within the plasmid, were used. In all cases, digestion was assayed by electrophoresis on an aliquot of the restriction digest on a 1% agarose gel, with a single band indicating linearity. The restriction digests were extracted and ethanol precipitated. The dried pellets were resuspended in $0.01/M$ Tris·HCl pH8, $0.001\ M$ EDTA at 1 μg/μl.

Transcription reactions were carried out in a volume of 25 μl, containing 40 mM Tris HCl, pH 7.5, 6 mM MgCl$_2$, 2 mM spermidine, 2 mM DTT, 1 μl of RNasin (Promega Biotech), 500 μm of all but the labeled ribotriphosphate, 5 μM of ^{35}S-UTP (800–1000 Ci/mmol; New England Nuclear) supplemented with cold UTP to 15 to 25 μM, 1 μl of linearized DNA template (=1 μg) plus 5 to 10 units of SP6 (Bethesda Research Laboratories) or T7 (Vector Cloning Systems) RNA polymerase. The reaction mix was incubated for 1.5 to 2 hours at 37° C. Incorporation was measured by TCA precipitation, with values of 40 to 70% incorporation typical, yielding 75 to 200 ng of RNA probe. The size of RNA probes was originally determined by electrophoresis under denaturing conditions through a 1.25-agarose glyoxyl gel, followed by exposure to x-ray film. In subsequent transcriptions, size was checked by agarose gel electrophoresis without denaturation. Under these conditions, using limited concentrations of ^{35}S-UTP, it is possible to obtain probe specific activities of greater than 1 × 10⁹ cpm/μg.

Table 8-1 Synopsis of *in situ* hybridization procedure using single-stranded RNA probes

---

1. Cryotomy
   a. *Fresh tissue.* After dissection, brain is placed in plastic vial, frozen in liquid nitrogen, and stored at −70° C. Ten 30-μm-thick sections are collected on acetylated Denhardt's-subbed slides and stored at −70° C.
   b. *Perfused tissue.* Perfusion with 4% buffered formaldehyde, followed by 18% sucrose infiltration. Ten 30-μm-thick sections are mounted on chrom-alum-subbed slides and stored at 4° C until they are used.

2. Tissue Pretreatment
   a. *Fixation.* 4% buffered formaldehyde for 20 minutes at room temperature (RT). Rinse two to three times in PBS.
   b. *Deproteination.* Place slides in predigested protease (0.25 mg/ml in 5 × TE) for 7.5 minutes at RT. Rinse well in PBS containing glycine (2 mg/ml) to block protease. Immerse slides in 0.02 N HCl for 10 minutes at RT. Rinse in PBS.
   c. *Postfixation.* 4% buffered paraformaldehyde for exactly 5 minutes at RT.
   d. *Dehydration.* In graded alcohols containing 0.33 M ammonium acetate, followed by air drying.

3. Prehybridization[a]
1 ml of prehybridization buffer containing 50% formamide, 5 × PIPES buffer with 0.75 M NaCl, 5 × Denhardt's, 0.2% SDS, and 250 μg/ml RNA and DNA is applied to the slides, which are then placed in sealed, humidified chambers for 2 to 3 hours at 45 to 50° C.

4. Hybridization
75 μl of hybridization buffer, containing 5 to 10 ng probe, is applied to the slides, which are sealed under coverslips for overnight hybridization at 45 to 50° C. Royalbond GRIP contact cement (Indal Aluminum Products, Los Angeles, CA) is used to provide a good temporary seal for the coverslips.

5. Posthybridization Rinses
   a. 4 × SSC for removal of coverslips.
   b. Pancreatic RNase digestion. 20 to 40 μg/ml in 0.5 M NaCl, 1 × TE for 30 minutes at 37° C. Follow with incubation in the same buffer without RNase for 30 minutes at 37° C.
   c. 2 × SSC for 30 minutes at RT, followed by a 0.1 × SSC rinse at elevated temperature.

---

[a]When using ^{35}S-labeled probes, DTT should be included in all steps from prehybridization through the first 4 × SSC rinse to prevent oxidation.

In Situ *Hybridization to Brain mRNAs Using RNA Probes*

A synopsis of our *in situ* hybridization protocol is shown in Table 8-1. This procedure works well with both perfused and fresh tissue, and with a variety of types and sizes of nucleic acid probes. The basic pretreatment and hybridization parameters have been adapted from a number of published methods (Hudson et al., 1981; Gee and Roberts, 1983; Brahic and Haase, 1978; Cox et al., 1984) and appear to provide a balance between allowing penetration of the probe to the mRNA of interest and subsequent retention of the hybrid in the tissue section. For posthybridization rinses, we have used the conditions outlined in Chapter 3.

Subtractive cDNA Hybridization for Isolation of Regionally Expressed mRNAs

We have used cDNA subtractive hybridization to isolate specific mRNA sequences expressed in one region of brain, for example, hippocampus, but

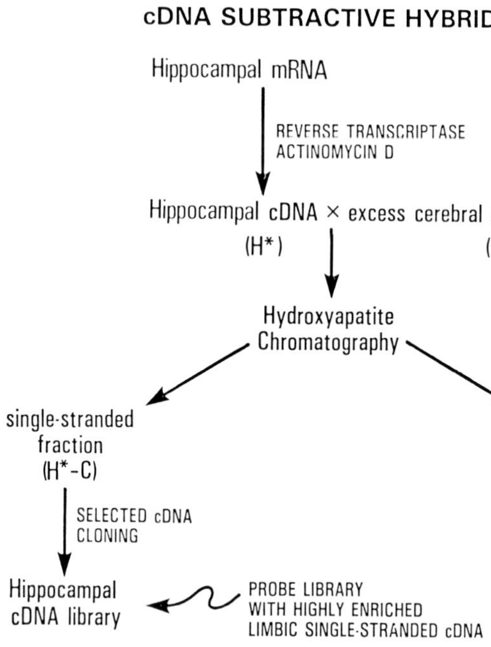

cDNA SUBTRACTIVE HYBRIDIZATION

Hippocampal mRNA

REVERSE TRANSCRIPTASE
ACTINOMYCIN D

Hippocampal cDNA × excess cerebral or cerebellar mRNA
(H*)   (C)

Hydroxyapatite
Chromatography

single-stranded   double-stranded
fraction   fraction
(H*-C)   (H* +C hybrids)

SELECTED cDNA
CLONING

Hippocampal   PROBE LIBRARY
cDNA library   WITH HIGHLY ENRICHED
                                       LIMBIC SINGLE-STRANDED cDNA

Figure 8-1 Strategy for isolation of subtracted cDNA. In the scheme presented here, subtractive hybridization was used for the isolation of hippocampal-directed cDNAs. Labeled cDNA was synthesized from hippocampal poly-A RNA, hybridized to a 10- to 20-fold excess of poly-A RNA from cerebellum or cerebral cortex, and separated into single-stranded (containing hippocampal-directed cDNA) and double-stranded nucleic acid on hydroxyapatite. In this case, the sscDNA was made double-stranded, cloned into pBR322, and the resultant colonies were screened with a twice-subtracted, hippocampal-specific probe (see Davis et al., 1984).

not other regions such as cerebellum or cerebral cortex. The experimental approach we have used, based on that of Davis et al. (1984), is shown in Figure 8-1.

Single-stranded (ss-) cDNA was prepared for hippocampal poly-A RNA in the presence of actinomycin D to prevent the self and exogenous primed synthesis of the second strand. The RNA template was removed by alkaline hydrolysis and purified by G50 chromatography, yielding 0.5 to 1.0 μg of cDNA. Hippocampal ss-cDNA, labeled with ^{32}p-dCTP, was hybridized to a 10- to 20-fold mass excess of poly-A RNA isolated from cerebellum or cerebral cortex. The hybridization mix was then applied to a 1.5-ml bed of hydroxyapatite, and the ss-cDNA, hippocampal-directed fraction eluted in a low-salt solution and reduced temperature. After subtracting cerebellar RNA sequences from hippocampal cDNA, this single-stranded fraction was found to contain 20 to 22% of the original mass of labeled input cDNA. Subsequently, the hippocampal-directed, ss-cDNA was rendered double-stranded and cloned into the Pst site of pBR322. Following identification by colony hybridization, individual cDNAs were isolated, labeled by nick translation and hybridized to RNA blots containing size-fractionated poly-A RNA from several brain regions, whole mouse brain, and liver. A more detailed presentation of this procedure and results will be presented in a subsequent publication (Higgins and Wilson, in preparation).

RESULTS

Methodological Considerations

Fixation

To determine if fixation by perfusion might provide better retention of RNA, as well as enhancing tissue morphology, we compared *in situ* hybridization to tissue sections from animals perfused with 4% buffered paraformaldehyde with fresh tissue, which was treated with the same fixative after cryotomy. The results of these experiments, using a single-stranded probe complementary to vasopressin mRNA, are shown in Figure 8-2. Both perfused and nonperfused fixed tissues displayed roughly similar intensities of specific hybridization. However, perfused tissues showed slightly higher backgrounds than sections of nonperfused tissue. Thus, although fixation by perfusion provides better preservation of morphological integrity, as can be judged from Nissl counterstaining of emulsion dipped sections, and should certainly be considered if immunocytochemistry is to be performed on the same or alternate sections, it may lead to slightly elevated background. This result has been reproduced with a variety of different probes, including RNA probes homologous to MuBr 2, 8, and 1B236 mRNAs (Higgins, Schmale, Bloom,

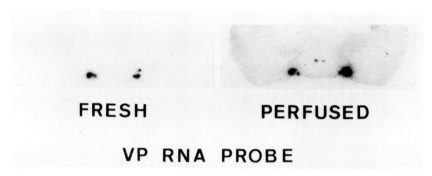

FRESH   PERFUSED

VP RNA PROBE

Figure 8-2 Comparison of *in situ* hybridization to fresh, postfixed versus perfusion-fixed tissue. This figure shows x-ray film images of coronal sections through rat hypothalamus treated in parallel under identical hybridization conditions. Note the elevated background signal present in tissue from the perfusion-fixed animal. *In situ* hybridization was performed as outlined in Table 8-1, using a single-stranded, ^{35}S-labeled RNA probe complementary to vasopressin mRNA (Schmale and Richter, 1984).

Wilson, and Milner, submitted) as well as double-stranded cDNA probes and synthetic oligonucleotide probes (unpublished observations).

Controls

Many experimental controls are possible including the following: (1) RNase pretreatment of tissue sections to show that loss of RNA eliminates hybridization signal (Branks and Wilson, 1986); (2) saturation of target mRNA sequence by hybridization with excess unlabeled probe prior to labeled probe hybridization to show that elimination of the specific mRNAs eliminates signal; (3) similarly, incubation of the labeled probe with excess unlabeled sense strand prior to or during hybridization; (4) hybridization with sense or nonsense probe to show that specific hybridization pattern is dependent on specific sequence of antisense probe; and (5) removal of specific hybridization signal by exceeding the stringency of the T_m range of the hybrid. Because of the specific nature of the hybridization between complementary nucleic acid strands, and the control made possible by removal of nonhybrid-bound RNA probe by pancreatic RNase treatment, however, it is apparent that *in situ* hybridization is not as dependent on controls as peptide immunohistochemistry or ligand/receptor autoradiography, for example. Appearance of possible technical artifacts (see the following), and the relative infancy of the application of this method to CNS tissue, however, requires some attempt to control for false-positive results. When using a new probe sequence, it is therefore important to define the extent of specific hybridization with the use of at least the previously listed controls 1 and 4. Again, the use of plasmids with dual RNA polymerase promoter sequences such as the pGEMINI vectors (Promega Biotech) facilitates the easy generation of sense-strand probes for control purposes (see Figure 8-3).

Artifacts

In our experience, several different false-positive images may result from *in situ* hybridization. Excluding those artifacts associated with emulsion autoradiography, we have observed a generalized neuronal, Nissl-like pattern that is not the result of specific hybrid formation. However, it should be emphasized that these artifactual patterns may be more common when double-stranded probes are used, and are not as evident when RNA probes are used for hybridization, where removal of nonhybrid-bound probe is facilitated by digestion with pancreatic RNase.

Figure 8-3a shows a characteristic Nissl-like artifactual image produced by hybridization with the sense strand (i.e., SP6 polymerase product) of pGEM8. The CA fields and dentate gyrus show up more strongly than the

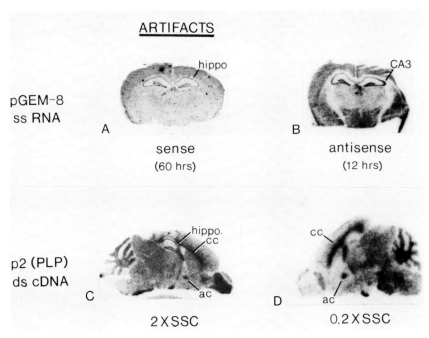

Figure 8-3 Artifacts associated with *in situ* hybridization to CNS tissue. These x-ray film images show several of the more common false-positive images that we have obtained with *in situ* hybridization. (A) A coronal section through mouse forebrain, showing nonspecific signals in CA regions and dentate gyrus of hippocampus (hippo) and pyriform cortex following hybridization using the sense strand of pGEM8. (B) Specific hybridization signal at a similar level of hypothalamus obtained with antisense product of pGEM8. Note the increased expression of MuBr 8 mRNA present within the CA3 field of the hippocampus as opposed to other CA regions and dentate gyrus. (C, D) Sagittal sections of mouse brain hybridized with a double-stranded cDNA probe complementary to proteolipid protein mRNA, showing a combination of specific hybridization, present within myelinated fiber tracts including anterior commissure (ac) and corpus callosum (cc), and the nonspecific signals shown in (C), indicating adherence of probe to neurons in regions such as the hippocampus, which were obtained under reduced stringency conditions (2 × SSC).

rest of the nonspecific background. In this case, the tissue was exposed to x-ray film (DuPont Cronex) for 60 hours. However, using the antisense strand, complementary to the MuBr 8 mRNA, three- to fourfold less exposure produces the specific image seen in Figure 8-3b. The nonspecific effects seems to be correlated with the density of RNA within different brain regions, with those areas showing up in Nissl-like stained material producing the strongest signal—that is, granule cell layer of the cerebellum, CA fields and dentate gyrus of the hippocampus, primary olfactory cortex, habenula, and olfactory bulb.

Occasionally, when using specific double-stranded probes at low hybridization and rinse stringency, it is possible to observe a combination of specific hybridization of this nonspecific RNA mass-associated phenomenon. Figure 8-3c shows a sagittal section of mouse brain hybridized with a nick-translated cDNA complementary to MuBr 2 (i.e., proteolipoprotein mRNA), and the CA fields and granule cell layers of the cerebellum show positive hybridization signals. At higher stringency, the artifactual Nissl-like pattern disappears (Figure 8-3d).

Very occasionally, it is also possible to observe a faint myelinlike, nonspecific hybridization signal (data not shown). Appropriate dehydration and removal of lipids from the tissue sections eliminate this kind of artifact.

MuBr 8 MRNA: A Widely Distributed, Neuronal mRNA

To determine in more detail the neuroanatomical distribution of a "brain-specific" mRNA isolated in a previous study (Branks and Wilson, 1986), a single-stranded, ^{35}S-labeled RNA probe homologous to the 3' portion of MuBr 8 mRNA was hybridized to coronal sections through the mouse brain. An overview of the expression of the MuBr 8 message is evident following exposure of tissue to x-ray film (Figure 8-4). In general, MuBr 8 mRNA is present in many regions of cerebral cortex, striatum, pons, cerebellum, and the ventral horn of the spinal cord, and in the limbic areas including the septum, amygdala, hippocampus, several hypothalmic nuclei, anterior thalamus, and the cranial nerve nuclei in the brain stem (Figure 8-4). Even at the gross level of resolution provided by x-ray film, several cytoarchitectonic features of the distribution of this mRNA can be observed. For example, in the hippocampus the transcript is more abundant in the CA3 field than in other CA regions (Figure 8-4 d and e), a feature that is even more apparent when the pattern obtained with the antisense probe is compared with that of the sense probe hybridization (Figure 8-3). Other nuclei showing high levels of expression include the ventromedial hypothalamic nucleus (Figure 8-4c), pontine nuclei (Figure 8-4f), and the ventral horn of the spinal cord.

To obtain cellular resolution, sections were dipped in emulsion, exposed for 3 to 5 days, and developed for autoradiography. Figure 8-5 shows the cellular expression of this mRNA in the hippocampal formation, where the differential abundance of the mRNA within the CA fields is even more ob-

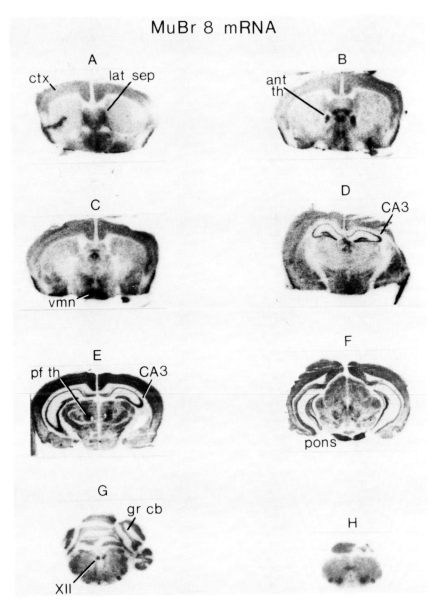

Figure 8-4 Overview of the distribution of MuBr 8 mRNA within mouse brain. X-ray film images of coronal sections from rostral (A) to caudal (H) levels of the CNS, showing the wide but specific expression of the mRNA within various neuronal cell groups. (Abbreviations: ant th, anterior thalamus; CA3, CA3 field of the hippocampal formation; ctx, cerebral cortex; gr cb, granule cell layer of the cerebellum; pf th, parafasicular nucleus of the thalamus; vmn, ventromedial hypothalamic nucleus; XII, hypoglossal nucleus.)

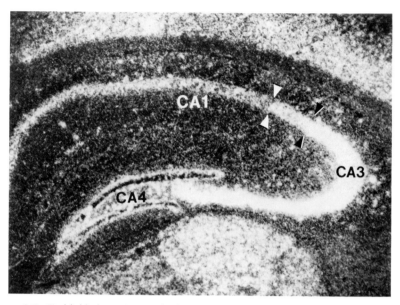

Figure 8-5 Darkfield photomicrograph showing the distribution of MuBr 8 mRNA within a coronal section of the murine hippocampal formation following *in situ* hybridization with a ^{35}S-labeled, single-stranded RNA probe. Slight overexposure emphasizes the differences in relative abundance of the mRNA between CA4, CA3, and other CA fields of the hippocampus, and the dentate gyrus, which appears devoid of MuBr 8 mRNA. A gradient of expression may be evident between the high level of expression in the CA3 intermediate levels, which may coincide with CA2 (*between black-and-white arrowheads*) and reduced levels in CA1. Note also the presence of MuBr 8 mRNA in overlying cerebral cortex, and ventrally, within the thalamus (Magnification, 10×).

vious than in our previous investigation using a nick-translated cDNA probe for *in situ* hybridization (Branks and Wilson, 1986). In fact, a gradient of expression can be observed within the CA fields, with the highest levels present in CA3, intermediate amounts in presumptive CA2 areas, and lowest levels in CA1 (Figure 8-5). In cerebral cortex, MuBr 8 appears to be present in cells throughout layers I through VI, and few differences in the level of expression can be observed between functionally or cytoarchitectonically distinct regions. For example, at the level of the rhinal fissure, where the transition occurs between neocortex dorsally and pyriform cortex ventrally, it is apparent that neurons throughout each of these cortical regions express the mRNA (Figure 8-6a). Other nuclei exhibiting high levels of expression of MuBr 8 mRNA include the anterior nuclear group of the thalamus (Figure 8-6b) and the lateral reticular nucleus (Figure 8-7).

Isolation of Regionally Expressed Transcripts by cDNA Subtractive Hybridization: Two mRNAs Shared by the Hippocampus and Amygdala

Subtractive cDNA hybridization was used to isolate mRNAs whose expression was shared between the hippocampal formation and the amygdaloid

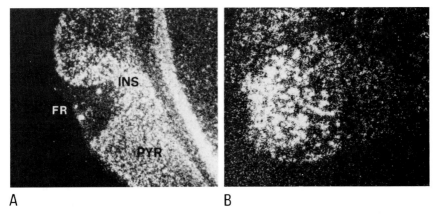

Figure 8-6   (A) Darkfield photomicrographs showing the uniform distribution of MuBr 8 mRNA within different layers of various cortical regions. Coronal sections: INS—insular cortex. PYR—pyriform cortex, FR—rhinal fissure (Magnification 10×). (B) Darkfield photomicrograph showing abundant expression of MuBr 8 mRNA within the anterodorsal thalamic nucleus; horizontal section (Magnification, 25×).

complex. Figure 8-8 shows the size and distribution of two hippocampal-directed mRNAs, as determined by RNA blotting. H-CbE3 (hippocampus-cerebellum, No. E3) mRNA is 4.3 kb in length, and is expressed in great abundance in hippocampus and amygdala, is expressed in lower abundance in the cerebral cortex, hypothalamus, and tectum, and is not present in cer-

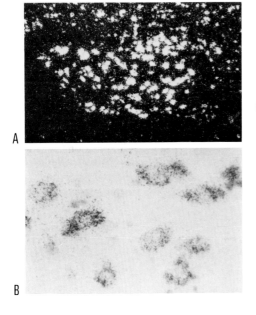

Figure 8-7   Localization of MuBr mRNA within neurons of the lateral reticular nucleus. (A) Darkfield photomicrograph showing uniform hybridization to cells of the lateral reticular nucleus; sagittal section (Magnification 25×). (B) Brightfield photomicrograph at a higher magnification of lateral reticular neurons showing cytoplasmic location of silver grains, indicating the presence of MuBr 8 mRNA. Note the cellular resolution obtainable with ^{35}S-labeled probes (Magnification, 158×).

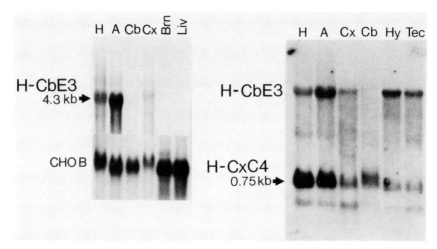

Figure 8-8 Examples of two mRNAs identified by hippocampal-directed subtracted cDNAs: shared expression in hippocampus and amygdala. RNA blots showing hybridization of H-CbE3 (*left and right blots*) and H-CxC4 (*right blot*) subtracted cDNA probes to poly-A RNA from hippocampus (H), amygdala (A), cerebellum (Cb), cerebral cortex (Cx), hypothalamus (Hy), tectum (Tec), whole brain (Brn), and liver (Liv). The relative abundance of these transcripts in different brain regions was determined by comparison of hybridization to a "housekeeping" RNA, CHO B. Approximately 5 μg of poly-A RNA was loaded in each lane, fractionated on 1.25% denaturing agarose gels, blotted onto nitrocellulose, and hybridized with ^{32}P-labeled cDNA inserts.

ebellum. In contrast, H-CxC4 (hippocampus-cortex, No. C4) mRNA is a small, 750-nt transcript that is present at high levels in the hippocampus and amygdala but can be found at low levels in a number of other brain regions (Figure 8-8). Currently, we are mapping the distribution of these mRNAs using *in situ* hybridization to determine the cellular sites of expression of these novel limbic gene products.

DISCUSSION

In the present study we have demonstrated an approach by which the neuroanatomical distribution of novel gene products isolated by cDNA cloning can be determined directly by *in situ* hybridization at an early step in characterization of the mRNA. The recent development of sensitive *in situ* hybridization methods, coupled with advances in cDNA cloning technology, provides a rapid and efficient strategy for mapping the expression of novel brain genes.

We previously showed that several brain-specific mRNAs isolated from a whole mouse brain cDNA library, although widely distributed in the mouse CNS, can be localized to limited cell populations within given brain regions such as hippocampus and cerebellum (Branks and Wilson, 1986). One of these mRNAs, which we have termed "MuBr 8," is a 2.2-kb mRNA transcript that encodes a protein without apparent homology to any known gene product (data not shown). To extend our preliminary studies on the anatomical distribution of this mRNA, which were based on limited analysis of *in situ* hybridization to sagittal brain sections using double-stranded cDNA probes, we have analyzed a more complete series of coronal sections hybridized to a single-stranded RNA probe. A number of neuronal cell groups have been shown to express to MuBr 8 mRNA at elevated levels. These include CA3 and CA4 pyramidal cells in the hippocampal formation, pontine nuclei, lateral reticular nucleus, anterior thalamus, cerebral cortex, and nuclei, which are more apparent on coronal sections, for example, the ventromedial nucleus of the hypothalamus, parafascicular thalamic nucleus, hypoglossal nucleus, and ventral horn of the spinal cord.

A more complete anatomical description of the cellular sites of expression of MuBr 8 does not allow a simple determination of the functional nature of the encoded protein. However, several functionally related cell groups preferentially express the mRNA. For example, MuBr 8 mRNA is abundant in structures involved in somatic motor function, including somatic motor cortex, pontine nuclei, and ventral horn of the spinal cord. In contrast, it is present in limbic areas such as the bed nucleus of the stria terminalis, hippocampus, several hypothalmic nuclei, and anterior thalamus. Although the presumption that the molecular identity of this gene product can be gathered from analysis of its systematic distribution seems artificial, the isolation of markers for specific cell types or populations provide useful tools to study developmental events such as cell migration and differentiation. For example, MuBr 8 mRNA is specifically expressed in abundance by CA3 and CA4 pyramical neurons and might, therefore, provide a useful marker for identifying these cells during development.

MuBr 8 mRNA belongs to a class of transcripts that are expressed across many different brain areas (Branks and Wilson, 1986; Milner and Sutcliffe, 1983; Sutcliffe et al., 1983). To isolate mRNAs that might be expressed in a more restricted regional distribution in brain, and therefore might provide more useful markers for specific cell types such as in the hippocampal formation, we have used subtractive cDNA hybridization and recombinant cDNA cloning of these sequences to isolate mRNA sequences expressed in hippocampus but not in other brain regions such as the cerebral cortex or cerebellum. Successful isolation of mRNAs expressed in the hippocampus and amygdala but not in the areas that they were subtracted from, as shown by RNA blotting (see Figure 8-8), demonstrates that this approach may be generally useful for the isolation of gene products differentially represented between neuroanatomical structures. Similarly, specific mRNA sequences expressed at different times during development, or under different behavioral or physiological states, may be isolated by these means.

In the strategy used by Sutcliffe et al. (1983), antibodies to synthetic peptides predicted from the nucleotide sequence of cDNA clones were used for immunocytochemical localization of novel brain proteins. Although this approach does provide useful information about the axonal and dendritic ramifications of cells expressing a novel protein not obtainable with *in situ* hybridization, it requires relatively complete cDNA sequence for determination of potential open reading frames and is dependent on the immunogenic properties of synthesized peptides. Also, the generation and characterization of antisera and questions of specificity associated with immunocytochemistry must be addressed. *In situ* hybridization is a more direct way of determining, at an early step of the screening process, the neuroanatomical distribution of cell bodies expressing a particular, uncharacterized mRNA. It can also provide information not apparent from immunocytochemical studies. For example, *in situ* hybridization of 1B236 mRNA has revealed its transient expression by cells in white matter areas of postnatal day 20 rat brain. This was not as apparent using antisera to synthetic 1B236 peptides, because fiber staining may have obscured cell body immunoreactivity (see Chapter 9).

 In situ hybridization can be used in a variety of ways to study gene and peptide expression in the CNS. We have shown how it can be applied to screen the anatomical distribution of newly discovered brain genes and have shown examples of technical artifacts that should be considered when mapping novel mRNAs. In its most straightforward and simplest application, *in situ* hybridization can be used to localize mRNAs encoding known brain peptides (Uhl et al., 1985; Gee et al., 1983), although in this regard, it may not provide more information than can be gathered from immunocytochemical studies. It may also be used to determine the expression of specific mRNAs prior to immunocytochemical localization of their translation products in developing CNS. Finally, *in situ* hybridization may be used to study regulatory events in gene expression such as determination of differential splicing events through the use of exon-specific probes (Amara et al., 1982; see also Chapter 9), differential promoter utilization (Shaw et al., 1985), and analysis of the expression of related mRNAs transcribed from divergent but related gene families, with synthetic oligonucleotides differing by single- or multiple-base changes (Connor et al., 1983; Wood et al., 1985).

ACKNOWLEDGMENTS

We thank M. Dietrich and P. Graber for help in the preparation of the manuscript. This work was supported by NIH grant No. CA09256 and by an NIH postdoctoral fellowship, NRSA No. NS07528, to G.A.H. This is publication No. 4307MB from the Research Institute of Scripps Clinic.

REFERENCES

Amara, S.G., Jonas, V., Rosenfield, M.G., Ong, E.S., and Evans, R.M. Alternative RNA processing in calcitonin gene expression generates mRNAs encoding different polypeptide products. *Nature* 298:240–244, 1982.

Bantle, J.A. and Hahn, W.E. Complexity and characterization of polyadenylated RNA in the mouse brain. *Cell* 8:139–150, 1976.

Bloom, F.E.,. Battenberg, E.L.F., Milner, R.J., and Sutcliffe, J.G. Immunocytochemical mapping of 1B236, a brain-specified neuronal polypeptide deduced from the sequence of a cloned RNA. *J. Neurosci.* 5(7):1781–1802, 1985.

Brahic, M. and Haase, A.T. Detection of viral sequences of low reiteration frequency by *in situ* hybridization. *Proc. Natl. Acad. Sci. USA* 75(12):6125–6129, 1978.

Branks, P. and Wilson, M.C. Patterns of gene expression in the murine brain revealed by *in situ* hybridization of brain specific messenger RNAs. *Mol. Brain Res.* 1:1–19, 1986.

Chaudhari, N. and Hahn, W.E. Genetic expression in the developing brain. *Science* 220:924–928, 1983.

Chikaraishi, D.M. Complexity of cytoplasmic polyadenylated and nonpolyadenylated rat brain ribonucleic acids. *Biochemistry* 18:3249–3256, 1979.

Conner, B.J., Reyes, A.A., Morin, C.M., Itakua, K., Teplitz, R.L., and Wallace, R.B. Detection of sickel cell β^s-globin allele by hybridization with synthetic oligonucleotides. *Proc. Natl. Acad. Sci. USA* 80:278–282, 1983.

Cox, K.H., DeLeon, D.V., Angerer, L.M., and Angerer, R.C. Detection of mRNAs in sea urchin embryos by *in situ* hybridization using asymmetric RNA probes. *Dev. Biol.* 101:485–502, 1984.

Davis, M.M., Cohen, D.I., Nielsen, E.A., Steinmetz, M., Paul, W.E., and Hood, L. Cell-type-specific cDNA probes and the murine I region: The localization and orientation of A^d. *Proc. Natl. Acad. Sci. USA* 81:2194–2198, 1984.

Emson, P.C. (Ed.). *Chemical Neuroanatomy.* New York: Raven Press, 1983.

Gee, C.E., Chen, C.C.-L., Roberts, J.L., Thompson, R., and Watson, S.J. Identification of proopiomelanocortin neurones in rat hypothalamus by *in situ* cDNA–mRNA hybridization. *Nature* 306:374–376, 1983.

Gee, C.E. and Roberts, J.L. *In situ* hybridization histochemistry: A technique for the study of gene expression in single cells. *DNA* 2:155–161, 1983.

Hockfield, S. and McKay, R.D.G. Identification of major cell classes in the developing mammalian nervous system. *J. Neurosci.* 5(12):3310–3328, 1985.

Hudson, P., Penschow, J., Shine, J., Ryan, G., Niall, H., and Coghlan, J. Hybridization histochemistry: Use of recombinant DNA as a "homing probe" for tissue localization of specific mRNA populations. *Endocrinol.* 108:353–356, 1981.

Levitt, P. A monoclonal antibody to limbic system neurons. *Science* 223:299–301, 1984.

Milner, R.J. and Sutcliffe, J.G. Gene expression in rat brain. *Nucleic Acids Res.* 11:5497–5520, 1983.

Schmale, H. and Richter, D. Single base deletion in the vasopressin gene is the cause of diabetes insipidus in Brattleboro rats. *Nature* 308:705–709, 1984.

Shaw, P., Sordat, B., and Schibler, U. The two promotors of the mouse α-amylase gene *amy*-1[a] are differentially activated during parotid gland differentiation. *Cell* 40:907–912, 1985.

Sternberger, L.A., Harwell, L.W., and Sternberger, N.H. Neurotypy: Regional individuality in rat brain detected by immunocytochemistry with monoclonal antibodies. *Proc. Natl. Acad. Sci. USA* 79:1326–1330, 1982.

Sutcliffe, J.G., Milner, R.J., Shinnick, T.M., and Bloom, F.E. Identifying the pro-

162 IN SITU HYBRIDIZATION

tein products of brain-specific genes with antibodies to chemically synthesized peptides. *Cell* 33:671–682, 1983.

Uhl, G.R., Zingg, H.H., and Habener, J.F. Vasopressin mRNA *in situ* hybridization: Localization and regulation studied with oligonucleotide cDNA probes in normal and Brattleboro rat hypothalamus. *Proc. Natl. Acad. Sci. USA* 82:5555–5559, 1985.

Wood, W.I., Gitschier, J., Lasky, L.A., and Lawn, R.M. Base composition-independent hybridization in tetramethylammonium chloride: A method for oligonucleotide screening of highly complex gene libraries. *Proc. Natl. Acad. Sci. USA* 82:1585–1588, 1985.

9

Localization of Cell-Type–Specific mRNAs by *In Situ* Hybridization

Robert J. Milner, Gerald A. Higgins,
Hartwig Schmale, and Floyd E. Bloom

The mammalian nervous system expresses more genes than any other tissue. Measurements of RNA complexity, which provides an estimate of gene activity, indicate that the complexity of the mRNA population of the rodent brain is severalfold greater than that of nonneural tissues such as liver or kidney (Bantle and Hahn, 1976; Chaudhari and Hahn, 1983). Our experiments suggest that approximately 30,000 different genes may be expressed in the brain of an adult rat and that more than half of these genes may be expressed predominantly or specifically in the brain (Milner and Sutcliffe, 1983). The extent and tissue specificity of gene expression in the nervous system is perhaps not surprising, given the morphological diversity of neuronal and nonneuronal cell types and the diversity of the functions exhibited by these cells (Bloom, 1984). The proteins that determine the morphology and the particular functions of neural cells are likely to be encoded by genes that may be specifically or predominantly expressed in neural tissue.

The mammalian nervous system therefore expresses a complex population of mRNAs that are distributed across a highly heterogeneous cell population. Not all of these genes, however, may be expressed in all brain cells. Genes that are expressed only in particular subpopulations of neural cells may make a significant contribution to the total gene activity of the nervous system. We can therefore expect to find genes that mark particular cell populations. This is clearly the case for the expression of one functional class of neural molecules, the neurotransmitters; anatomical studies over the past two decades have exhaustively documented the cell populations in which particular neurotransmitters and neuropeptides are expressed singly or in combination with other transmitters (Bloom, 1984). Studies using monoclonal antibodies (reviewed in Valentino et al., 1985) have shown how molecules other than transmitters may also mark particular cell types. A major goal of modern neuroanatomy is to identify and exploit such markers of neural cell type to provide a basis for understanding the morphological and functional differences between neural cells.

Recombinant DNA technology provides the means to characterize molecules rapidly and accurately at the level of nucleic acid. For example, several neural molecules originally identified at the protein chemical level have been characterized further by molecular cloning of their mRNAs: tyrosine hydroxylase (Grima et al., 1985), myelin basic protein (Roach et al., 1983; Zeller et al., 1984) glial fibrillary acidic protein (Lewis et al., 1984), neurofilament protein (Lewis and Cowan, 1985), S-100 protein (Kuwano et al., 1984), and the protein precursors for many neuropeptides (reviewed by Douglass et al., 1984). Each of these proteins can be regarded as cell-type-specific markers. For example, myelin basic protein is expressed by oligodendrocytes during and after myelination, and glial fibrillary acidic protein is expressed only by astrocytes, whereas tyrosine hydroxylase and neuropeptide precursors are each expressed by subpopulations of neurons.

We have used recombinant DNA techniques to identify brain proteins that may provide additional cell type markers by selecting cDNA clones of rat brain mRNAs that are expressed specifically or predominantly in the brain (Milner and Sutcliffe, 1983; Sutcliffe et al., 1983; Milner et al., in press). Clones that hybridized to brain mRNAs but not to liver or kidney mRNAs were selected from a cDNA library prepared from adult rat brain cytoplasmic poly-A RNA library. Approximately 30% of the clones examined fell into this category, and we have operationally defined the mRNAs corresponding to these clones as "brain specific." These studies have resulted in descriptions of the brain-specific protein 1B236, which is expressed only in a particular population of neurons (Sutcliffe et al., 1983; Malfroy et al., 1985; Bloom et al., 1985; Lenoir et al., 1986); rat brain proteolipid protein, which is expressed only in oligodendrocytes (Milner et al., 1985); and the identifier (ID) sequence, a repetitive genetic element that may regulate brain gene expression (Sutcliffe et al., 1984).

THE BRAIN-SPECIFIC GENE 1B236

The brain-specific gene 1B236 has served as the model for our first extensive analysis of brain gene expression by *in situ* hybridization. To facilitate understanding of our results with this gene, a brief review of the 1B236 system may be instructive. The clone p1B236 hybridizes to an mRNA of 2500 nucleotides that is present in brain with an abundance of approximately 0.01% but is undetectable in liver or kidney (Sutcliffe et al., 1983). A second 1B236 mRNA, 3000 nucleotides in length, is also detectable in rat tissues in lower abundance. Both 1B236 mRNAs show a heterogeneous but parallel distribution in RNA preparations from different regions of rat brain, with the highest concentrations in the thalamus, midbrain, and pons-medulla (Malfroy et al., 1986). The nucleotide sequence of p1B236 provided the 318 amino acid carboxy terminal sequence of the corresponding putative protein (1B236), which had no homologies to any previously defined sequence as determined by computer search. To demonstrate the existence of this protein in the brain and to characterize its properties, three peptides (P5, P6, and P7) were synthesized, corresponding to nonoverlapping regions of the translated protein sequence (Figure 9-1), and were used to raise antipeptide antibodies in rabbits.

In immunocytochemical experiments, antisera against each of the three peptides produced identical cytological maps of neuronal staining in adult rat brain; the reactivity was blocked in each case by preincubation of the antibodies with the appropriate peptide (Sutcliffe et al., 1983; Bloom et al., 1985). The fact that antibodies against each of three nonoverlapping regions of the 1B236 sequence give similar staining patterns provides strong evidence that these antibodies react with the protein encoded by the clone p1B236, and therefore that this protein does exist in the rat brain. The immunoreactivity was distributed heterogeneously in the brain and was most pronounced within olfactory bulb and peduncle, specific hypothalamic and preoptic nuclei, the neostriatum, limbic and neocortical regions, particularly somatosensory cortex (Figure 9-2). The data suggest that 1B236 is present in particular neuronal elements within functionally related central systems and that they may have evolved from related progenitor lines (Bloom et al., 1985).

The predominant form of the 1B236 protein in the brain appears to be a 100,000-dalton glycoprotein, which is probably membrane associated (Malfroy et al., 1985). In addition, peptide fragments of the 1B236 protein can be detected in brain extracts, using the same antipeptide antibodies. We had postulated previously that the 1B236 protein might be proteolytically processed to generate a number of different peptides (Sutcliffe et al., 1983), based on structural similarities between the 1B236 protein and hormone or neuropeptide precursors (Douglass et al., 1984). The multiplicity of 1B236 immunoreactive forms suggests that this molecule undergoes extensive posttranslational modification.

During brain development in the rat, 1B236 mRNA is first detectable in extracts of whole rat brain at 5 days after birth (PD5) and increases to a maximum concentration at PD25 (Lenoir et al., 1986). In extracts of dissected brain regions, 1B236 mRNA is first detectable at PD5 in hindbrain and cerebellum, at PD9 in midbrain, but not until PD13 in telencephalon. The appearance of 1B236 protein follows a very similar time course to that of its mRNA in both the whole brain and dissected brain regions, suggesting that the expression of the protein is regulated largely by transcription of its mRNA. The pattern of 1B236 expression was confirmed by immunocytochemical localization of 1B236 protein: immunoreactive material can be detected first in the spinal cord at PD3–PD5, then appears in progressively more rostal brain regions in increasingly older animals, occuring last in cerebral cortex. Several brain regions, however, that do not contain 1B236 immunoreactivity in the adult, such as optic nerve and somatic efferent cranial nerve nuclei, show transient expression of 1B236 during postnatal de-

1B236

LysSerThrGlyGluAspAsnArgAlaThrValMetThrAlaProThrLysProThrValAlaSerValMetThrAlaProThrLysProThrValAlaSerVal...

Protein/DNA sequence block (1B236 cDNA insert):

```
                        LysSerThrGlyGluAspAsnArgAlaThrValMetThrAlaProThrLysProThrValAlaSerValMetThrAlaProThrLysProThrValAlaSerVal Se
CTGCA(G)₁₂AGAAGTCCTATGGCCAGGACAACCGCACGGTGAGCTGAGCTCATGTATGCACCTTGAAGCCCACAGTGAATGGACGGTGTGGCGTAGAGGGGAGCAGTTC

rIleLeuCysSerThrAsnSerAsnProAspProIleLeuThrIlePheLysGluLysGlnIleLeuLeuIleThrValIleTyrGluSerGlnLeuGluGluProAlaValThrP
CATCCTGTTGTTCCACACAGAGCAACCCGGACCTATTCTCACCATCTTCAAGGAGAAGCAGATCCTGCCACGGTCATCTATGAGAGTCAGCTGGAACTCCTGCAGTGACGC
                                                                                                    Pst I

roGluAspAspGlyGluTyrTrpCysValAlaGluAsnGlnTyrGlyArgAlaThrAlaThrAlaProPheAsnLeuSerValAlaProPheIleLeuLeuSerHisCysAlaAla
CCGAGGACGATGGGGAGTACTGGTGTGTAGCTGAGAACCAGTATGGCCAGAGACCACGGCTTCAACCTGTCTGTGGAGTTTGCCCATAATCCTTCTGGAATCGCACTGTCGAGCG

AlaArgAspAspThrValGlnCysValValLysSerAsnProPheSerValValAlaProPheGlnLeuProSerProSerArgAsnValThrGluLeuArgGluValPheTyrSe
GCCAGAGACGACCGTGCAGTGCGTCGTGAAATCCAACCCGTTCAGCTGCCTTTGAGCTGCCTTCCCCAACGTGACTGGAGCAGAGAGAGGAGTTTGTACTC

rGluGluProSerIleLeuThrLeuArgGlyGlnAlaGlnAlaAlaProProArgValIleCysThrSerArgAsnLeuCysTyrArgGlnThrGlnSerLeuGluLeuProP
AGAGCCAGAGCCGGCCTCTGCTCACCAGCATCCTCACGCTCCGGGGTCAGGCCCAGGCCGCACCCCCGCGTCATTTGTACCTCCAGGAACCTCACGCCCAGAGCCTGGCCTT

heGlnGluAlaLaHisArgLeuMetThrAlaLysIleGlyProValGlyAlaValValAlaValAlaPheAlaAlaLeuAlaLeuValCysTyrIleThrGlnThrArgArgLysAsnVal
TCCAGGGAGCACGACGACTGATGTGGGCCAAAATCGGCCTGTGGGTGCTGTGGTCGCCTTTGCCATCCTGATTGCCATTGTCTGCTACATCACCCAGACAAGAAAAAAAAGAACGTC
                                                                                                             P5

ThrGluGluProSerPheSerAlaGlyGluAspAsnProHisValLeuThrSerProAspGluAlaIleSerGlnGluAlaLysGlyGlySerGluAspArgAr
ACAGAGGAGCCCAGCTTCTCAGCGGGAGACAACCCTCATGTCCTGACAAGTCCTGAGCCACCTGATAAGTATGAAGTGAAGAGCGCCTGGGGTCCGAGAGGAG
                                                                   Eco RI                                                      P7

gLeuLeuGluLeuArgGlyGluProProProLeuLeuProProGlnSerLeuAspSerLeuSerLysGlyLysArgProThrLeuAspSerThrLeuThrGluGlaLeuValGlyThrAlaG
GCTGCTGGGCCTTAGGGGAGGAACCCCCAGAACTGGACCCAGTTATTCCCACTCAGACCTGGGAAACGAGCCACCCTGACAGAGGAGCTGGCTCAGTACGCAG

*IleSerThrTrpIle***
AAATCCGACCACATGAGGAAGCTGGGGCTGGCCCTGTGGCTCACCCCCCATCAGGACCCTCGCTTGGCCCCACTGGCCGTGGGCTCCTTTCTTCTTTGAGAGTGAGTAGGGGTGGGGG

CGGGAAGGGGCGGGCAGGAGAAACAGTGAGGTCTTAGGGCCCGGCCTCCCCTCCTTCTGCCAACATCCTGCACCTATGTTACAGCTCCTCCCCTCCTTTTA

ACCTCAGCTGTTGAGAGGGGTGCTCTGTCTGTCCATGTATGTTATTGTTATCCTGGTCTGTCCTGTCCTGTCCTGTCCTGGTCCCAGGACCTGTACAAAAGGGACATGAAATAAATGTCCTAAT

GACAAGTGCCAGTCTAGACCCATCCTTTGGAGGAAAGGGGCATATTAGTAATACTTTTCTGTTGCATTGTCTAACAAAATACTGGACAAAAACAC(A)₋₁₀₀(C)ₙ
```

Figure 9-1 Nucleotide sequence of the cDNA insert of the clone p1B236, showing the translated protein sequence and the positions of synthetic peptides (P5, P6, P7, underlined) used for the preparation of antibodies in rabbits. The locations of sites for the restriction endonucleases Pst I and Eco RI are also shown: the sequence between these sites was subcloned to provide a probe for 1B236 mRNA. (Modified from Sutcliffe et al., 1983.)

1B236 IMMUNOREACTIVE NEURONS

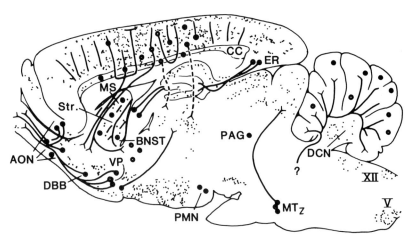

Figure 9-2 Neuroanatomical distribution of 1B236 in the adult rat brain: schematic overview of neuropil and perikaryonal immunoreactivity patterns. Cell-body rich areas are symbolized by circles, neuropil-rich regions by dots; presumptive fiber trajectories are symbolized by solid lines and putative pathways by dotted lines. (Abbreviations: AON, anterior olfactory nucleus; BNST, bed nucleus of the stria terminalis; CC, corpus callosum; DBB, diagonal band of Broca; DCN, deep cerebellar nuclei; ER, entorhinal cortex; MS, medial septal nucleus; MTz, medial trapezoid nucleus; PAG, periaqueductal gray; PMN, premammillary nucleus; Str, striatum; V, spinal trigeminal nucleus; VP, ventral pallidal areas; XII, hypoglossal nucleus. Taken from Bloom et al., 1985.)

velopment. The postnatal expression of 1B236 indicates that this protein is a marker for the terminal differentiation of particular neurons and suggests that the 1B236 protein probably mediates functions specific to the adult nervous system.

In the adult rat brain, 1B236 antipeptide antibodies are predominantly reactive with fiber systems. To visualize the cell bodies that synthesize the 1B236 protein with our antipeptide antibodies, it was necessary either to treat adult animals with colchicine to inhibit axonal transport of 1B236 (Bloom et al., 1985) or to investigate the expression of 1B236 in developing animals, in which sufficient newly synthesized material is still present in the cell bodies (Lenoir et al., 1986). Neither approach is perfect: colchicine is highly toxic and results in a considerable pertubation of normal neuronal morphology; in the study of developing animals we must take into account possible age-dependent differences in the expression of 1B236. Furthermore, in either case, cell bodies within fiber tracts were particularly difficult to resolve by immunocytochemistry when the surrounding fibers were highly immunoreactive for 1B236. For these reasons we felt it necessary to reexamine the expression of the 1B236 gene with an alternative technique. The procedure of *in situ* hybridization provided an appropriate approach.

The technique of *in situ* hybridization extends the application of molecular biology directly to neuroanatomical analysis, as this volume bears witness. In our laboratories this technique has found two main uses: (1) in improving and verifying our understanding of the neuroanatomical distribution of molecules uncovered by our molecular approaches and (2) in selecting clones of cell-type-specific mRNAs by virtue of their heterogeneous distribution in the brain. These approaches are illustrated here by our studies on the localization of mRNAs for the precursor of the peptide hormone arginine vasopressin (AVP) and for the brain-specific protein 1B236.

METHODS

Clone-Construction

Clones of AVP precursor mRNA (Schmale and Richter, 1984) and 1B236 mRNA (p1B236, Sutcliffe et al., 1983) were digested with appropriate restriction endonucleases and single fragments were subcloned in the polylinker of vectors pSP64 and pSP65 (Promega Biotec, Madison, WI). The positions of the Pst I and Eco RI sites used to generate a fragment of p1B236 are shown in Figure 9-1.

Transcription of RNA Probes

Both plasmids were linearized with Hind III, which cuts at a single site in the polylinker on the 3' side of the inserted fragment in each case. Transcription reactions incorporating either ^{32}P-GTP or ^{35}S-UTP were carried out as described in Chapter 8.

In situ Hybridization

The methods used for the experiments are described completely in Chapter 8. Hybridized tissue sections were exposed first to x-ray film and subsequently dipped in photographic emulsion and developed for autoradiography.

RESULTS

Localization of Vasopressin mRNA

To develop *in situ* hybridization techniques within our laboratories, we chose a model system—the peptide hormone AVP— which has a well-characterized localization in particular neurons in the rodent brain. *In situ* hybridization experiments on brain sections using either nick-translated DNA probes or RNA probes synthesized and labeled with ^{32}P or ^{35}S demonstrate a discrete distribution of AVP precursor mRNA. Autoradiographs of brain sections by contact exposure of x-ray film show two major concentrations of hybridized AVP probe in the supraoptic and paraventricular nuclei (Figure 9-3) of the hypothalamus. Additional signals of lower intensity can be seen in other areas, including lateral hypothalamus and the suprachiasmatic nucleus (Figure 9-3). Autoradiographic studies allow the visualization of AVP mRNA within individual magnocellular neurons in the paraventricular and supraoptic nuclei (Figure 9-4). This distribution is consistent with the known immunocytochemical distribution of AVP (Bujis, 1980; Hou-Yu et al., 1982; Vandesande et al., 1975) and with the localization of AVP precursor mRNA by others (Uhl et al., 1985; Fuller et al., 1985; Nojiri et al., 1985). The accurate localization of AVP precursor mRNA, with low background and no apparent artifacts, indicated that our *in situ* hybridization technique was appropriate to tackle the distribution of other, less well-defined neural mRNAs.

Localization of 1B236 mRNA in the Adult Brain

For the studies of 1B236 gene expression, a 557-nucleotide Pst I to Eco RI fragment of the clone p1B236 was subcloned in the RNA transcription vector pSP65. This fragment corresponds to part of the coding region for the 1B236 protein (see Figure 9-1); the fragment is oriented within the vector so that transcription of the plasmid by SP6 polymerase produces an antisense

VASOPRESSIN mRNA: IN SITU HYBRIDIZATION

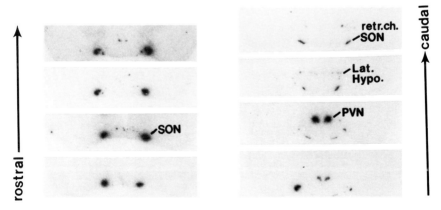

Figure 9-3 *In situ* hybridization of vasopressin mRNA within hypothalamic nuclei. Coronal rat brain sections were hybridized with a ^{32}P-labeled RNA probe complementary to AVP precursor mRNA and exposed to x-ray film for 36 hours. (SON, supraoptic nuclei; PVN, paraventricular nucleus; Lat.Hypo., lateral hypothalamus; ret.ch.SON, retrochiasmatic portion of the supraoptic nucleus.)

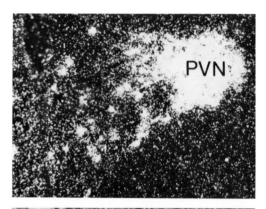

Figure 9-4 Cellular localization of vasopressin mRNA. Tissue sections were hybridized with a ^{35}S-labeled RNA probe, dipped in photographic emulsion, exposed for 7 days, and developed for autoradiography. The high-magnification darkfield photomicrographs show silver grains over vasopressin mRNA containing magnocellular neurons in the paraventricular (PVN, *top*) and supraoptic (SON, *bottom*) hypothalamic nuclei.

strand to the 1B236 mRNA. This probe, labeled with ^{35}S, was used for all the experiments described here.

In the adult brain the localization of 1B236 cell bodies by *in situ* hybridization was found to be consistent with their localization by immunocytochemistry in untreated animals or in animals treated with colchicine. Thus, brain sections exposed to x-ray film showed intense labeling of areas corresponding to olfactory bulb and accessory olfactory nuclei, cortical regions, particularly cingulate, somatosensory and primary olfactory cortex, and the granule cell layer of the cerebellum (Figure 9-5). We previously showed 1B236 immunoreactive cell bodies in each of these regions by immunocytochemistry after colchicine treatment (see Figure 9-2; Bloom et al., 1985). The same brain sections that had been dipped in emulsion showed concentrations of silver grains over presumptive neurons in each of these regions, confirming the cellular localization of the 1B236 mRNA probe (data not shown).

Localization of 1B236 mRNA in the Developing Brain

We continued our studies of 1B236 expression using brains from 20-day-old (PD20) rat pups. There were several reasons for this experimental choice.

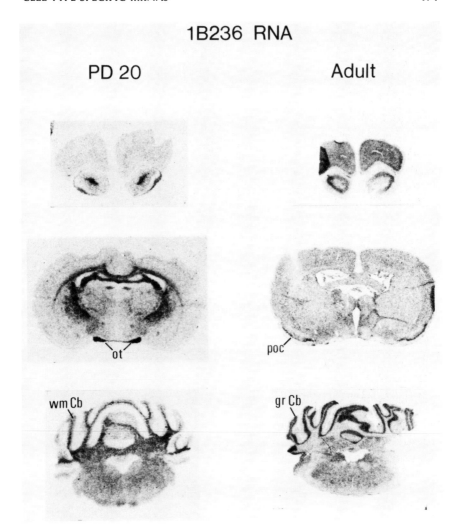

Figure 9-5 *In situ* hybridization of 1B236 mRNA in adult and PD20 rat brains. Tissue sections were hybridized with a ^{32}P RNA probe complementary to 1B236 mRNA and exposed to x-ray film for 36 hours. Matched coronal sections are shown at the level of the rostral telecephalon (*top*), caudate–hypothalamus (*center*), and medulla–cerebellum (*bottom*).

First, we had previously demonstrated that the concentration of 1B236 mRNA in the whole brain was approximately threefold higher at this age than in adults (Lenoir et al., 1986), possibly allowing easier detection of 1B236 mRNA. Second, this analysis had indicated that the adult pattern of 1B236 expression unfolded during the second and third weeks of postnatal development and that all the adult systems were in place by PD20. Third, we have used our analysis of the developing brain to confirm cell body assignments in the adult brain. *In situ* hybridization provided the means of verifying these assignments independently.

The distribution of cells expressing 1B236 mRNA in the PD20 brain is quite different from that of an adult. The distinctive and striking differences in the expression of 1B236 at the two ages can be seen in representing sections taken from three different levels of the neuroaxis (see Figure 9-5). We have undertaken a detailed analysis of the distribution of 1B236 cells at PD20 (Higgins, Schmale, Bloom, Milner, and Wilson, in preparation), in order to compare this with that of the adult and our previous developmental results. When possible we have localized 1B236 protein on alternate sections by immunocytochemistry, using the antipeptide antibodies; these usually showed a similar concentration of cell bodies within the same regions expressing 1B236 mRNA.

The differences between the expression of 1B236 mRNA at PD20 and in the adult focus largely on its expression in white matter regions of the developing animal (Table 9-1). At PD20 the highest numbers of cell bodies appear to be concentrated within myelinated fiber tracts and bundles. For example, cells expressing 1B236 mRNA are found in all parts of the anterior commissure (Figure 9-6a), corpus callosum (Figure 9-6b), optic tract, internal capsule, and fornix. Additional cells were observed in the caudate-putamen in a patchlike pattern corresponding to striate fiber bundles. In the cerebellum, cells expressing 1B236 mRNA appear to be concentrated largely within white matter areas, as is clearly visible at the level of the x-ray film image (see Figure 9-5). The densest concentration of positive cells in the brain stem are found within the pyramidal tract (Figure 9-6c); 1B236 immunoreactive cells can be distinguished within the pyramidal tract, whose fibers are themselves intensely immunoreactive for 1B236.

We previously described a transient expression of 1B236 immunoreactivity in optic tract and cranial nerve nuclei during postnatal development (Lenoir et al., 1986). The expression of 1B236 mRNA in white matter shows a

Table 9-1 Changes in expression of 1B236 mRNA during postnatal development

Structure	PD 20	PD 100
White matter		
Anterior commissure	+++	+
Corpus callosum	+++	+
Fornix	+++	−
Stria medullaris	+++	−
Stria terminalis	+++	−
Mammillothalamic tract	+++	−
Fasciculus retroflexus	+++	−
Pyramidal tract	+++	−
Internal capsule	+++	−
External capsule	+++	−
Cerebellar white matter	+++	−
Optic tract	+++	−
Gray Matter		
Golgi cells (cerebellum)	−	++
Primary olfactory cortex	+	+++

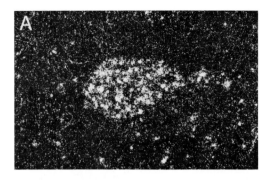

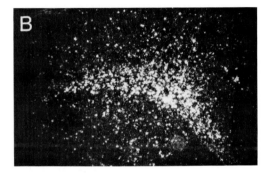

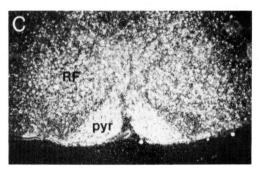

Figure 9-6 Cellular localization of 1B236 mRNA. Tissue sections were hybridized with a ^{35}S-labeled RNA probe, dipped in photographic emulsion, exposed for 7 days, and developed for autoradiography. The darkfield photomicrographs show silver grains located over clusters of cells in coronal sections through (A) anterior commissure, (B) corpus callosum, and (C) brain stem (pyr, pyramidal tract; RF, reticular formation).

similar transience: in many structures, such as the pyramidal tract and the cerebellar white matter, no 1B236-positive cells are detectable in the adult by either immunocytochemistry or *in situ* hybridization. Other white matter areas, however, such as anterior commissure and corpus callosum, retain 1B236-containing cells in the adult, although in reduced numbers.

In contrast, in gray matter areas at PD20, 1B236 mRNA was found within neuronal cell groups that also express 1B236 mRNA in the adult. For example, there is extensive 1B236 expression in the olfactory system at both ages. In the olfactory bulb, the highest number of cells expressing 1B236 mRNA are located within the internal plexiform and granular layers. In the olfactory peduncle 1B236 mRNA is found within polymorphic cells of the olfactory tubercle, the posterior part of the accessory olfactory nucleus and primary olfactory cortex. Positive cells are found within other regions of neocortex, particularly layers V and VI of somatosensory cortex. Scattered cells are also present within the lateral division of the bed nucleus of the stria terminalis and within the basolateral and lateral nuclei of the amygdala. There are few, if any, cells containing 1B236 mRNA within the hippocampus and dentate gyrus, in contrast to their high concentration in adjacent white matter areas. Positive cells are found in the lateral geniculate nucleus and stria medullaris but not within neuron-rich areas of the habenula. Several brain-stem nuclei contain 1B236-positive cells, including the medial trapezoid nucleus, the dorsal tegmentum, and the dorsal motor nucleus. In adult animals, all of these gray matter areas contain cells expressing 1B236, shown by *in situ* hybridization or immunocytochemistry (Bloom et al., 1985). The postnatal expression of 1B236 mRNA in gray matter areas, therefore reflects the development of adult neuronal systems; the number of 1B236-positive cells within neuronal nuclei remains relatively constant with development.

DISCUSSION

The results presented here demonstrate the validity of *in situ* hybridization as a neuroanatomical tool, as is amply demonstrated by the other chapters in this volume. For two different neural systems—AVP and the protein 1B236—we have demonstrated that the localization of cell bodies expressing the particular mRNA is consistent with previous immunocytochemical studies. In addition, the developmental analysis of 1B236 by *in situ* hybridization has revealed the transient expression of 1B236 in populations of presumptive neurons located within myelinated tracts during late postnatal development.

The technique for *in situ* hybridization that we have used in these studies and appropriate controls are described more fully in Chapter 8. Our best procedure uses RNA probes labeled with ^{35}S. As described by Cox and coworkers (1984; see also Chapter 3), these probes can be generated at high enough specific activities to be sufficient for autoradiographic detection either by exposing them to x-ray film or by directly dipping slides into photographic emulsion, with relatively short exposure times in most cases.

The localization of AVP mRNA to magnocellular cells in the paraventricular and supraoptic nuclei and adjacent areas of the hypothalamus (Figures 9-3 and 9-4) is entirely consistent with the results of previous studies on the distribution of AVP synthesizing cells by immunocytochemistry (Bujis, 1980; Hou-Yu et al., 1982; Vandesande et al., 1975) and of recent studies on AVP mRNA localization using *in situ* hybridization (Uhl et al., 1985; Fuller et al., 1985; Nojiri et al., 1985). The distribution of cells expressing 1B236 mRNA in the adult rat, demonstrated by *in situ* hybridization, is also consistent with our previous study results using immunocytochemistry after treatment with colchicine (Bloom et al., 1985). In both cases, *in situ* hybridization provided an entirely independent experimental approach for the verification of the sites of expression of particular genes.

In situ hybridization proved its worth as a neuroanatomical tool in our studies of the developmental expression of 1B236, however, by revealing the transient expression of 1B236 mRNA in a population of cells in fiber tracts and other white matter areas. Our previous studies showed an intense but transient expression of 1B236 protein, largely in fibers, within these same areas during postnatal development (Lenoir et al., 1986). Although some 1B236 immunoreactive cell bodies could also be detected in these areas, cell bodies within the fiber tracts were often difficult to distinguish or define because of the overwhelming immunoreactivity of the surrounding neuropil. In contrast, 1B236 cells were strikingly displayed by *in situ* hybridization because of the ability of this technique to visualize only cell bodies. Furthermore, even in the developing animal, in whom cell bodies can be visualized by immunocytochemistry without colchicine treatment, most 1B236 immunoreactivity is found in fibers, either singly or in larger tracts. Therefore, global or regional assessments of gene expression based on immunocytochemistry analysis can easily be biased by the contribution of fiber staining relative to that of cell body staining. *In situ* hybridization, however, provides a focus solely for cell bodies. Although this may be an advantage in some situations, it is gained at the expense of not visualizing axon fibers or even resolving details of cellular morphology that are critical for distinguishing different cell types. In the future, perhaps, complete anatomical investigations of cell type markers will involve the complementary advantages of both *in situ* hybridization and immunocytochemistry.

The expression of 1B236 mRNA and protein in cells and fibers within white matter areas is largely transient: a few positive cells can be detected in the adult rat brain within the anterior commissure and corpus callosum, but not in other white matter regions. Our earlier studies showed that 1B236 immunoreactivity in optic nerve and pyramidal tract was first detectable at PD9 and had virtually disappeared by PD20. The expression of 1B236 is therefore concurrent with the myelination of these fiber tracts. In fact, for the whole brain, the expression of 1B236 mRNA during development (Lenoir et al., 1986) coincides very closely with the time course of expression of the mRNA for proteolipid protein, a major component of myelin (Milner et al., 1985). It is unlikely that oligodendrocytes are directly responsible for the transient expression of 1B236, because many areas known to contain oligodendrocytes do not show 1B236 hybridization. On the other hand, the 1B236 gene may be expressed by neurons within these tracts in response to myelination; or it may be expressed by a novel population of cells that serve some accessory function during the myelination of large fiber tracts. Alternatively, the correspondence with myelination may be pure coincidence, and the timing of 1B236 gene expression may reflect some other process in the maturation of these fiber tracts.

Although the clone p1B236 was selected because it hybridized to a brain-specific mRNA, subsequent studies at the level of both RNA and protein have demonstrated that 1B236 is also a cell-type-specific marker, expressed only in particular populations of neural cells. With the possible exception of the transient white matter cells described here, however, it is not yet clear what these cells have in common other than the expression of 1B236. Nevertheless, the heterogeneous pattern of expression of 1B236 mRNA was revealed by *in situ* hybridization in adult or developing brains even at the level of the gross x-ray film image (Figure 9-5), suggesting that this technique might be useful for selecting further clones of cell-type-specific mRNAs.

We are currently using *in situ* hybridization to characterize cDNA clones corresponding to developmentally regulated mRNAs. Several possible clones have been selected from a cDNA library prepared from 16th-embryonic-day (ED16) brain poly-A RNA, using a subtracted cDNA probe (Hedrick et al., 1984), prepared by hybridizing ^{32}P-labeled ED16 cDNA with an excess of adult brain mRNA and removing double-stranded molecules. In preliminary experiments, at least one of these clones hybridized to discrete cell populations in patterns detectable in x-ray images (Miller, Naus, Lai, Higgins, and Milner, unpublished observations).

ACKNOWLEDGMENTS

We thank Elena Battenberg, Cary Lai, D. Lenoir, F. Miller, C. Naus, and Michael Wilson for discussions and advice. The studies reported in this chapter were supported by NIH grants NS 20728 and NS 21815, NIAAA Alcohol Research Center grant AA 06420, and McNeil Laboratories. This is publication BCR-4273 from the Research Institute of Scripps Clinic.

REFERENCES

Bantle, J.A. and Hahn, W.E. Complexity and characterization of polyadenylated RNA in the mouse brain. *Cell* 8:139–150, 1976.

Bloom, F.E. The functional significance of neurotransmitter diversity. *Am. J. Physiol.* 246:C184–C194, 1984.

Bloom, F.E., Battenberg, E.L.F., Milner, R.J., and Sutcliffe, J.G. Immunocytochemical mapping of 1B236, a brain-specified neuronal polypeptide deduced from the sequence of a cloned mRNA. *J. Neurosci.* 5:1781–1802, 1985.

Bujis, R.M. Immunocytochemical demonstration of vasopressin and oxytocin in the rat brain by light and electron microscopy. *J. Histochem. Cytochem.* 28:357–360, 1980.

Chaudhari, N. and Hahn, W.E. Genetic expression in the developing brain. *Science* 220:924–928, 1983.

Cox, K.H., DeLeon, D.V., Angerer, L.M., and Angerer, R.C. Detection of mRNAs in sea urchin embryos by *in situ* hybridization using asymmetric RNA probes. *Dev. Biol.* 101:485–502, 1984.

Douglass, J., Civelli, O., and Herbert, E. Polyprotein gene expression: Generation of diversity of neuroendocrine peptides. *Annu. Rev. Biochem.* 53:665–715, 1984.

Fuller, P.J., Clements, J.A., and Funder, J.W. Localization of arginine vasopressin-neurophysin II messenger ribonucleic acid in the hypothalamus of control and Brattleboro rats by hybridization histochemistry with a synthetic pentadeca-mer oligonucleotide probe. *Endocrinology* 116:2366–2368, 1985.

Grima, B., Lamouroux, A., Blanot, F., Biguet, N.F., and Mallet, J. Complete coding sequence of rat tyrosine hydroxylase mRNA. *Proc. Natl. Acad. Sci. USA* 82:617–621, 1985.

Hedrick, S.M., Cohen, D.L., Nielsen, E.A., and Davis, M.M. Isolation of cDNA clones encoding T cell-specific membrane-associated proteins. *Nature* 308:149–153, 1984.

Hou-Yu, A., Ehrlich, P., Valiquette, G., Engelhardt, D., Sawyer, W., Nilaver, G., and Zimmerman, E. A monoclonal antibody to vasopressin: Preparation, characterization, and application in immunocytochemistry. *J. Histochem. Cytochem.* 30:1249–1260, 1982.

Kuwano, R., Usui, H., Maeda, T., Fukui, T., Yamanari, N., Ohtsuka, E., Ikehara, M., and Takahashi, Y. Molecular cloning and the complete nucleotide sequence of cDNA and mRNA for S-100 protein of rat brain. *Nucleic Acids Res.* 12:7455–7465, 1984.

Lenoir, D., Battenberg, E.L.F., Kiel, M., Bloom, F.E., and Milner, R.J. The brain-specific gene 1B236 is expressed postnatally in the developing rat brain. *J. Neurosci.,* 6:522–530, 1986.

Lewis, S.A. and Cowan, N.J. Genetics, evolution and expression of the 68,000-mol-wt neurofilament protein: Isolation of a cloned cDNA probe. *J. Cell Biol.* 100:843–850, 1985.

Lewis, S.A., Balcarek, J.M., Krek, V., Shelanski, M., and Cowan, N.G. Sequence of a cDNA clone encoding mouse glial fibrillary acidic protein: Structural conservation of intermediate filaments. *Proc. Natl. Acad. Sci. USA* 81:2743–2746, 1984.

Loh, Y.P., Brownstein, M.J., and Gainer, H. Proteolysis in neuropeptide processing and other neural functions. *Annu. Rev. Neurosci.* 7:189–222, 1984.

Malfroy, B., Bakhit, C., Bloom, F.E., Sutcliffe, J.G., and Milner, R.J. Brain-specific polypeptide 1B236 exists in multiple molecular forms. *Proc. Natl. Acad. Sci. USA* 82:2009–2013, 1985.

Malfroy, B., Bakhit, C., Lenoir, D., and Milner, R.J. *J. Neurochem.* submitted, 1986.

Milner, R.J. and Sutcliffe, J.G. Gene expression in rat brain. *Nucleic Acids Res.* 11:5497–5520, 1983.

Milner, R.J., Lai, C., Nave, K.-A., Lenoir, D., Ogata, J., and Sutcliffe, J.G. Nucleotide sequences of two mRNAs for rat brain myelin proteolipid protein. *Cell* 42:931–939, 1985.

Milner, R.J., Bloom, F.E., and Sutcliffe, J.G. *Current Topics in Developmental Biology,* in press.

Nojiri, H., Sato, M., and Urano, A. *In situ* hybridization of the vasopressin mRNA in the rat hypothalamus by use of a synthetic oligoucleotide probe. *Neurosci. Lett.* 58:101–105, 1985.

Roach, A., Boylan, K., Horvath, S., Prusiner, S., and Hood, L.E. Characterization of cloned cDNA representing rat myelin basic protein: Absence of expression in brain of shiverer mutant mice. *Cell* 34:799–806, 1983.

Schmale, H. and Richter, D. Single base deletion in the vasopressin gene is the cause of diabetes insipidus in Brattleboro rats. *Nature* 308:705–709, 1984.

Sutcliffe, J.G., Milner, R.J., Shinnick, T.M., and Bloom, F.E. Identifying the protein products of brain-specific genes with antibodies to chemically synthesized peptides. *Cell* 33:671–682, 1983.

Sutcliffe, J.G., Milner, R.J., Gottesfeld, J.M., and Reynolds, W. Control of neuronal gene expression. *Science* 225:1308–1315, 1984.

Uhl, G.R., Zingg, H.H., and Habener, J.F. Vasopressin in mRNA *in situ* hybridization: Localization and regulation studied with oligonucleotide cDNA probes in normal and Brattleboro rat hypothalamus. *Proc. Natl. Acad. Sci. USA* 82:5555–5559, 1985.

Valentino, K.L., Winter, J., and Reichardt, L.F. Applications of monoclonal antibodies to neuroscience research. *Annu. Rev. Neurosci.* 8:199–232, 1985.

Vandesande, F., Dierickx, K., and DeMey, J. Identification of the vasopressin-neurophysin producing neurons of the rat suprachiasmatic nuclei. *Cell Tissue Res.* 156:377–380, 1975.

Zeller, N.K., Hunkeller, M.J., Campagnoni, A.T., Sprague, J., and Lazzarini, R.A. Characterization of mouse myelin basic protein messenger RNAs. *Proc. Natl. Acad. Sci. USA* 81:18–22, 1984.

10

Localization of Peptide Hormone Gene Expression in Adult and Embryonic Tissue

John E. Pintar and Delia Ines Lugo

In situ hybridization of cellular mRNA provides an exciting complement to immunocytochemistry. Identification of individual cells containing specific mRNA strongly suggests that the cells are actively engaged in the synthesis of specific proteins. Thus, endogenous synthesis of an immunodetectable protein can be distinguished from uptake of a protein synthesized elsewhere. Moreover, the usefulness of immunocytochemistry will be compromised if peptide levels are below the limits of detection, as may be the case during the earliest stages of cell differentiation, or if small peptides are lost from cells that lack a mature storage apparatus prior to immunocytochemical detection. *In situ* hybridization thus provides a direct means of both assessing the earliest times of gene expression during development and clarifying sites of synthesis for gene products in which immunologic demonstration has proven elusive or controversial.

In the procedures discussed here and in most other chapters in this volume, radiolabeled polynucleotides (either cDNA or cRNA) are incubated with tissue sections, annealed to complementary mRNA sequences in the section by specific base pairing, and subsequently detected by autoradiography. The success of localizing specific mRNA sequences within tissue sections principally depends on optimization of the following conditions: (1) maximal retention of cellular mRNA and maintenance of adequate morphology during fixation and preparation of tissue sections; (2) maximal accessibility of the target RNA for interaction with the corresponding hybridization probe; and (3) sensitive and specific probe binding and visualization.

In this chapter, we compare some of the variables influencing the sensitivity of *in situ* hybridization to cellular mRNA in sections of adult and embryonic mammalian tissues and emphasize our applications of these procedures thus far. Our studies have primarily used probes to pituitary peptide hormones (e.g., proopiomelanocortin [POMC] and prolactin [PRL], which have differential distributions in the various pituitary regions) and growth factors (e.g., nerve growth factor [NGF] and epidermal growth factor [EGF]), which are present in high abundance in the male mouse submaxillary gland. The abundance of mRNAs in these cells has made these systems especially suitable for examining some of the different factors affecting *in situ* hybridization to mammalian cellular mRNAs.

FIXATION AND TISSUE SECTION PREPARATION

Fixation

One of the most important requirements of the procedure is to maximally retain nondegraded mRNA in the tissue being examined. Retention of cellular RNA can be readily monitored by staining tissue sections with acridine orange throughout the procedure (e.g., see Hafen et al., 1983). To minimize the activity of endogenous and exogenous ribonucleases, sterile plasticware should be routinely used, gloves should be worn throughout the procedure, and 0.1% diethylpyrocarbonate (DEP; Shivers et al., 1986) should be added to fixatives and the sucrose solution used as a cryoprotector in cryostat embedding of fixed tissue.

A variety of fixatives (including those that precipitate macromolecules such as ethanol [EtOH] or EtOH/acetic acid) and those that can cross-link RNA to protein (including a variety of aldehyde fixatives; see Haase et al., 1984, for discussion) have been used successfully by different investigators. For example, EtOH and EtOH/acetic acid have been used to localize large (15-kb) viral genomes and viral RNA in tissue culture cells and paraffin sections of viral infected tissue (Brahic and Haase, 1978; Haase et al., 1984; see also Chapter 11), 1% glutaraldehyde combined with protease treatment has proven optimal for sea urchin embryos (Angerer and Angerer, 1981; Cox et al., 1984), while 4% paraformaldehyde has been extensively used for fixation of *Drosophila* embryos and a variety of mammalian tissues (Gee et al., 1983; Hafen et al., 1983; Pintar et al., 1984; McGinnis et al., 1984; Brulet et al., 1985; Wilcox et al., 1986; Shivers et al., 1986). Following formaldehyde fixation, intact POMC mRNA has been demonstrated in RNA extracted from formaldehyde-fixed tissue (Kelsey et al., 1986). We have compared the *in situ* hybridization signals of pituitary and submaxillary gland tissue sections hybridized with peptide hormone and growth factor probes following fixation with 4% formaldehyde, pH 7, in PBS, 4% formaldehyde–0.5% glutaraldehyde in PBS, EtOH/acetic acid (3:1) and acetic acid/EtOH/chloroform (6:3:1; Carnoy's). Hybridization signals in both submaxillary glands and adult pituitary have been essentially equivalent following both EtOH/acetic acid and formaldehyde fixation but much reduced following Carnoy's fixation.

Aldehyde fixatives not only provide good morphological preservation but also are compatible with immunocytochemical detection of a wide range of antigens. Formaldehyde fixation, for example, has been used successfully to colocalize specific peptides and the corresponding mRNA in the same section (Griffin et al., 1983; Blum et al., 1984; Shivers et al., 1986; see Chapter 6).

Because many of our studies have also used immunocytochemistry, essentially all of our *in situ* hybridization studies of mammalian embryonic gene expression have used tissue fixed with formaldehyde before they were embedded and sectioned. Embryos are fixed with 4% formaldehyde (made fresh from paraformaldehyde) in PBS for 1 to 2 hours. A similar fixation procedure has detected very abundant repetitive sequences in early mouse embryos (Brulet et al., 1985). We routinely perfusion-fix rat embryos older than embryonic day 15 (e15) and have obtained stronger signals following hybridization of late fetal and early neonatal perfused pituitaries with cDNA peptide hormone probes compared with those of immersion-fixed tissue. It should be noted that formaldehyde postfixation of fresh-cut sections of frozen tissue may further increase the hybridization signal of mammalian tissues (Shivers et al., 1986; Fremeau and Roberts, personal communication; see also Chapter 6); we are presently exploring this possibility for mammalian embryonic and fetal tissue.

Embedding Procedures and Collection of Sections

Cryostat sections of frozen tissue and sections from paraffin and plastic-embedded tissue have been effectively used for *in situ* hybridization studies. In general, cryostat sections have been used when immunocytochemical detection of specific antigens is required (c.g., Gee et al., 1983), since this procedure retains antigenicity of a wider range of antigens than paraffin embedding protocols. On the other hand, paraffin sections have been used successfully to demonstrate abundant nervous-system-specific cDNAs in rat embryos (Anderson and Axel, 1985) and have routinely been used to study the extent of viral infection in specific tissues (Haase et al., 1984) as well as to detect moderately abundant mRNA in sea urchin embryos (Angerer and Angerer, 1981; Cox et al., 1984; see Chapter 3). Finally, methacrylate-embedded tissue, which provides superior morphology but is more difficult to section rapidly, has been used to demonstrate histone mRNA in *Xenopus* oocytes (Jamrich et al., 1984).

Paraffin-embedded tissue generally yields better morphologic preservation than frozen tissue and, if comparable in sensitivity to cryostat-prepared material for *in situ* hybridization studies, would be especially useful in allowing rapid processing and hybridization of young (e.g., <e10 rat) mammalian embryos.

We have directly compared the *in situ* hybridization signals with paraffin-embedded and cryostat sections of perfused adult rat pituitaries following hybridization with cRNA probes. The results of these studies (which used

[32]P-prolactin and ^{3}H-POMC cRNA probes) demonstrated that after otherwise identical conditions (which did not include a postfixation step after protease treatment), stronger signals were observed on cryostat sections than paraffin sections (Figure 10-1). In addition, in parallel experiments using more stringent washing conditions, signal intensity was further diminished significantly in paraffin-embedded tissues, while the signal intensity in frozen sections was unaffected. Similar results have been observed following hybridization of submaxillary gland tissue sections with growth factor probes (Gresik, personal communication). These results suggest that hybridized RNA is being lost from the paraffin-embedded sections during the wash procedures. It may be possible to eliminate this loss by postfixation of paraffin sections following protease treatment (Angerer and Angerer, 1981) or more rapid paraffin processing for small tissues. However, until these procedures are further explored, it appears that paraffin sections should be used cautiously for *in situ* hybridization experiments on mammalian tissue in which sensitivity is of utmost importance.

One of the most frustrating aspects of the *in situ* methodology has been loss of sections during the hybridization and wash procedures. Most of our early studies used gelatin chrom-alum subbed slides for hybridizations of

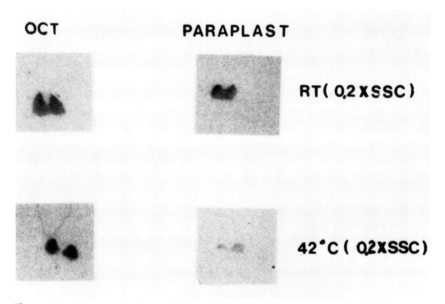

Figure 10-1   Comparison of paraffin- and OCT-embedded sections. Adult pituitaries were perfusion-fixed in formaldehyde-PBS and either processed for conventional paraffin embedding or equilibrated with sucrose and embedded in OCT. Paraffin and OCT sections were subsequently hybridized with cRNA probes to prolactin and proopiomelanocortin using the procedure outlined in Figure 10-4. Specific labeling of the anterior pituitary lobe following hybridization with ^{32}P-prolactin cRNA is seen on both paraffin and cryostat sections, but the signal is lower on the paraffin sections. This difference is more pronounced following washing at more stringent conditions. Similar results were observed following emulsion coating of sections hybridized with ^{3}H-POMC cRNA.

both cDNA and cRNA probes. In a slight modification of standard subbing protocols, we found that coating acid-washed slides for tissue section collection with 0.15% gelatin and 0.1% chromium potassium sulfate markedly increases the retention of sections. Dehydration of sections following protease treatment and preceding prehybridization has proven useful in preventing section fall-off and allows washing of gelatin-coated slides hybridized and washed at temperatures up to 45° C (gelatin melting temperature is about 52° C) with negligible loss of sections treated with 1 µg/ml proteinase K and 30 µg/ml RNase. Our recent studies have shown that poly-L-lysine–coated slides (Cox et al., 1984), which allow hybridization and stringent posthybridization washes of paraffin-embedded tissue sections at elevated temperatures for long periods of time, are also suitable for collecting cryostat sections of both fixed and frozen tissue; at present this may be the method of choice for slide preparation.

INTERACTION OF TARGET mRNA WITH CORRESPONDING HYBRIDIZATION PROBE

The accessibility of the target RNA and diffusion of probe into the section can generally be enhanced by treatments designed to remove proteins from the fixed tissue section or to reverse the cross-linking of protein and RNA. Further, there is evidence that adjusting the size of the hybridization probe can facilitate its penetration into the sections.

Section Treatment

Mild deproteination (either by mild acid treatment to remove basic proteins or by protease digestion of sections) generally enhances the *in situ* hybridization signal (Gee and Roberts, 1983; Brahic and Haase, 1978). Most investigators subject fixed tissue to protease treatment prior to *in situ* hybridization to render the tissue RNA more exposed and thus increase the probability of encountering and interacting with the probe. Several proteases have been used often in conjunction with acid treatment. Pronase has been found by some groups to be more effective than proteinase K (Singer and Ward, 1982; Hafen et al., 1983); whereas other groups have preferred proteinase-K (Cox et al., 1984; Angerer et al., 1985). We have observed increases in signal intensity of pituitary sections hybridized with peptide hormone probes following treatment with 1 µg/ml of proteinase K (which is available in a nuclease-free state and can be readily inactivated) and now routinely use this treatment, although in earlier studies we successfully used mild acid (0.2 N HCl for 10′) treatment. (Figure 10-2; see also Figure 10-5).

It should be emphasized that the extent of deproteination required appears to depend on the extent of fixation, although different data have been re-

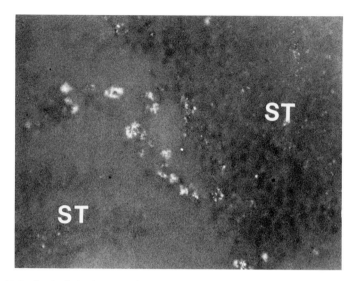

Figure 10-2 *In situ* hybridization of testicular tissue section. An adult male rat testis was immersion fixed for 2 hours at room temperature in freshly prepared 4% paraformaldehyde-phosphate–buffered saline (pH 7.0), equilibrated overnight in 20% sucrose, and embedded in OCT. Eight-micron cryostat sections were hydrated and treated with 0.2 *N* HCl for 10 minutes. Sections were then prehybridized at 24° C for 1 hour with prehybridization buffer containing 50% formamide, 4 × SSC, 20 mM Tris, pH 7.5, 0.1 mg/ml yeast total RNA, 0.05 mg/ml yeast transfer RNA, 0.1 mg/ml herring sperm DNA, 1 m*M* EDTA, 0.02% ficoll, 0.02% polyvinyl pyrrolidone-20, and 0.2 mg/ml BSA. Prehybridization buffer was removed and replaced with hybridization buffer containing the preceding components plus 10% dextran sulfate and ^{3}H-nick-translated DNA (specific activity 10^7 cpm/μg; 2×10^4 cpm/section). The hybridization mix was heat-denatured at 100° C for 10 minutes before being applied to the tissue section. Incubation of tissue sections with hybridization mix continued overnight at 24° C. Sections were washed for 1 hour each in two changes of 2 × SSC containing 0.1% SDS at room temperature and for 5 hours in 0.2 × SSC containing 0.1% SDS at 37° C. Sections were air dried and coated with Ilford L-4 emulsion that had been diluted 1:1 with 0.6 *M* ammonium acetate. Sections were then exposed at 4° C for 6 weeks, developed with D-19, and counterstained with hematoxylin/eosin. Silver grains were observed in the cytoplasm of most Leydig cells in the interstitial space of the testis; cells within the seminiferous tubules (ST) are unlabeled.

ported for different tissues. For example, maximal enhancement of viral genome hybridization occurs following both acid and protease treatment of acid-alcohol-fixed tissue. On the other hand, mild protease treatment leads to loss of mRNA following a brief formaldehyde fixation of tissue culture cells (Lawrence and Singer, 1985; see also Chapter 4), whereas a similar protease treatment enhances the signal from glutaraldehyde-fixed sea urchin embryos (Angerer and Angerer, 1981; Cox et al., 1984; see also Chapter 3). Moreover, since different tissue types may have different requirements for deproteination, this procedure should be optimized for each type of tissue used. A number of investigators advocate postfixation with 4% paraformaldehyde after protease treatment of tissues to obtain better signals (Singer and Ward, 1982; Hafen et al., 1983). In our hands, such postfixation following proteinase K digestion has not enhanced signal detection of peptide hormone mRNA of adult rat pituitary cryostat sections.

Probe Length

Another factor that affects the extent of hybridization is the size of the hybridization probe. In general, shorter probes have provided stronger signals than long probes (Gee and Roberts, 1983; Hafen et al., 1983; Cox et al., 1984; Shivers et al., 1986). Nick-translation conditions for radiolabeling of double-stranded probes can be readily adjusted to produce fragments of average size 100 bases, which we have routinely used (Figure 10-2, Plate 7a; see Figure 10-4). The size of RNA probes can also be adjusted by base hydrolysis after synthesis (Cox et al., 1984). It should be noted, however, that optimal sizes reported by different groups have varied somewhat, ranging from 50 base pairs (Gee and Roberts, 1983) to 200 base pairs (Hafen et al., 1983) to greater than 750 base pairs for cRNA probes (see Chapter 7). These differences may reflect differences in tissue preparation prior to hybridization that also influence the penetration of the probe or, possibly, the requirement of longer hybridization times for complete hybridization of longer probes (see Chapter 11), which may not have been achieved in some experiments.

SENSITIVITY AND SPECIFICITY OF PROBE BINDING

The rate and extent of hybrid formation in solution depends on clearly defined parameters: the amount of probe, its complexity and GC content, salt concentration, and temperature of hybridization. In tissue sections, however, empirical observations have shown that both the rate and extent of hybridization are longer than the predicted times in solution (see the discussion in Chapter 11). Our prehybridization and hybridization conditions for both cDNA and cRNA hybridizations have been based on those initially developed by others (Brahic and Haase, 1984; Gee et al., 1983; Cox et al., 1984), with minor modifications that are included in the appropriate figure legends. The following discussion presents our comparisons of single- and double-stranded probes for hybridization, the relative advantages of different methods for labeling probes, and controls used to assess the specificity of hybridization.

Double-Stranded cDNA Probes versus Single-Stranded Probes

Much of our early work, and that of many others, with *in situ* hybridization successfully used nick-translated double-stranded probes. Originally, it was thought that nick-translated probes could provide a high degree of sensitivity by forming networks of overlapping nucleotide stretches including probe hybridized to target RNAs. On the other hand, self-reassociation of double-stranded probes during hybridization could compete with, and prematurely terminate, the *in situ* reaction. Direct comparisons of hybridization by asymmetric and symmetric probes (Angerer and Angerer, 1981) suggested that asymmetric probes were significantly more efficient. Quantitative comparisons (Cox et al., 1984; see Chapter 3) have confirmed these initial observations and demonstrate up to eightfold higher signals with the use of asymmetric RNA probes than with cDNA double-stranded probes at saturating probe concentrations. The advantage of single-stranded probes has been confirmed in our laboratory. We were unable to demonstrate specific hybridization at the single-cell level in rat prenatal anterior pituitary lobe cells following long exposure times using double-stranded H³-labeled POMC cDNA probes, while such specific labeling has been observed using single-stranded RNA probes with exposure times as short as 1 week (see later).

Single-stranded RNA probes have a number of other advantages over double-stranded cDNA probes. Lower background levels can be achieved with single-stranded RNA probes because the sections can be subjected to RNase digestion following hybridization, thus degrading any nonspecifically bound probe. RNA–RNA hybrids also have the advantage of being more stable than DNA–RNA hybrids, which permits the use of higher wash temperatures to reduce nonspecific probe binding.

A variety of vectors containing RNA polymerase sites are now commercially available and allow transcription of RNAs of opposite orientation; thus, one strand can serve as a control for hybridization specificity. Large quantities of labeled RNA of high-specific activity and constant length are readily produced, and subcloning of adjacent cRNA fragments into these vectors allows the production of nonoverlapping RNAs that can be hybridized to adjacent sections to ensure specificity of hybridization signals.

Labeling of Hybridization Probes

Both radioactive nucleotides and nonradioactive nucleotides to which biotin or other "reporter" molecules are attached can be incorporated into the hybridization probes. Although nonradioactively labeled nucleotides have been used successfully to detect abundant mRNA in cultured cells (Singer and Ward, 1982; Brigati et al., 1983; see also Chapter 4), the sensitivity of hybridization with these probes has not yet approached that of radiolabeled probes in experimental systems in which sensitivity can be readily determined (Haase et al., 1984). It seems likely, though, that multistep amplification procedures similar to those used for immunocytochemistry will enhance the sensitivity and usefulness of these probes.

A variety of commercially available radiolabeled nucleotides can be incorporated into hybridization probes, and each has advantages and disadvantages. ^{32}P is a high-energy β-emitter and can be used to quickly determine the regional distribution of a particular mRNA in tissue sections. Sections hybridized with ^{32}P-labeled probes can be directly exposed to x-ray film and have given information about relative distribution of pituitary peptide mRNA at different times of development (see following; Figure 10-3), but single-cell localization in a tissue containing closely apposed cells cannot be obtained following emulsion-coating of ^{32}P-hybridized tissue because of emission scattering.

Sections hybridized with ^{35}S-labeled probes can initially be examined by film autoradiography and subsequently be emulsion-coated to give adequate microscopic resolution. ^{35}S-labeled probes provide an approximately 10-fold increase in specific activity and a three- to fivefold increase in autoradiographic efficiency over ^{3}H, although background levels have generally been higher than with probes radiolabeled with other isotopes. Extensive washing of sections collected on polylysine-coated slides and hybridized with ^{35}S-labeled cRNA in buffer containing 1% sodium thiosulfate, 0.05% sodium pyrophosphate, and 0.1% β-mercaptoethanol at high temperature (55° C)

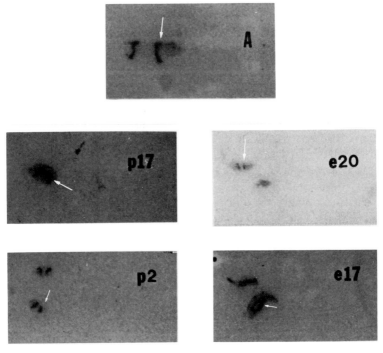

Figure 10-3 ^{32}P *in situ* hybridization of isolated rat pituitary glands. Embryos and neonates from e17 to postnatal day 17 and an adult were perfusion fixed with freshly made paraformaldehyde; the pituitaries were removed, cut at 10 μm, and hybridized with double-stranded POMC cDNA insert labeled with ^{32}P by nick translation using the protocol outlined in Figure 10-2. A higher density of POMC mRNA is present in the intermediate lobe (*denoted by arrows*) even as early as e17. (Note that unlabeled e20 and e17 blocks contained two pituitaries, one of which does not have intermediate lobe in the plane of section.)

substantially decreases the background. Single cells expressing peptide hormone mRNA in the brain and embryonic central nervous system cells expressing a specific homeo-box gene can be visualized after relatively short exposure times (Schachter, Fremeau, and Roberts, personal communication; Toth, Pintar, and Nguyen-Huu, in preparation). ^{35}S probes have proven useful for other studies as well (see Plate 7c; Lewis and Cowan, 1985; Anderson and Axel, 1985).

^{3}H-labeled probes have proved superior for single-cell resolution of tightly packed embryonic or adult cells that differ in gene expression (see Figure 10-2; Plate 7a; Figure 10-4). As a low-energy β-emitter, ^{3}H creates the least amount of scatter and correspondingly greater resolution; however, since the β-emission travels only 2 to 3 μm, hybridizing RNA deep within a section cannot be detected.

Optimal exposure times of emulsion-coated sections depend on the radioisotope used, the level of mRNA/cell, and the amount and specific activity of the probe used, and they must be determined empirically. Therefore, we always expose adjacent hybridized slides in different boxes so that the progress of the exposure process can be checked at different times.

Specificity of Probe Binding

The temperatures used for hybridization and posthybridization washes are important in determining the specificity of hybridization and improving the signal-to-noise ratio. Optimal hybridization and posthybridization treatments to maximize specific hybridization are discussed thoroughly in other chapters in this volume (see Chapter 3 and 11). Our procedures are based on those discussions and are reported in the figure legends.

A variety of controls can be performed to assess the specificity of the hybridization. The most direct control is to establish that the hybrid melts at the appropriate temperature (Gee and Roberts, 1983; Kelsey et al., 1986). Immunocytochemical detection of peptides or proteins encoded by the mRNA of interest on the same or adjacent sections has more generally been used to provide evidence for specific hybridization.

Further controls should be performed, especially during development of this technique: (1) Exposure of sections taken through the entire procedure, but not exposed to probe, should always be performed to eliminate the possibility of positive chemography, which is especially prevalent in the mammalian central nervous system (see Chapters 6 and 11). (2) Tissue sections can be digested with RNase prior to hybridization; this procedure should eliminate all signal and demonstrates that the signal results from binding of the probe to RNA in the tissue. (3) Hybridization of the section with unlabeled probe before radiolabeled probe is added should greatly decrease or eliminate signal. (4) The pattern of hybridizing cells should not be mimicked following hybridization with nonspecific probes of similar size and specific activity. (5) Identical hybridization patterns should be obtained with multiple nonhomologous probes hybridized to adjacent tissue sections (see later). This type of analysis (which is analogous to immunocytochemical detection with two monoclonal antibodies to different determinants of the same protein) is especially necessary when immunocytochemical localization has not suggested at least a potential synthetic site for specific proteins or peptides.

APPLICATIONS OF *IN SITU* HYBRIDIZATION

Demonstration of Gene Expression in Single Cells and Confirmation of Sites of Peptide Synthesis

The immunocytochemical detection of different peptide hormones in cell types not classically thought to be sites of synthesis for these hormones has provided many potential applications for *in situ* hybridization. For example, peptide hormones derived from the precursor hormone POMC have been immunocytochemically identified in the arcuate nucleus of the hypothalamus, which raised the question of whether or not these peptides are synthesized in these cells. This question is especially important because albumin and tranferrin, for example, can be concentrated by central nervous system neurons and demonstrated immunocytochemically even though synthesized elsewhere. For the POMC-peptide containing arcuate neurons, endogenous synthesis was confirmed by the *in situ* localization of POMC mRNA in the identical cells containing POMC peptide immunoreactivity (Gee et al., 1983).

In our initial application of *in situ* hybridization procedures, we approached a similar problem. POMC-derived peptides had surprisingly been localized immunocytochemically in Leydig cells of the adult rat testis. We used a nick-translated POMC insert to demonstrate that POMC mRNA is in fact present in these cells (Pintar et al., 1984; see Figure 10-2). Further analysis of the distribution of Leydig cells containing POMC mRNA has revealed that Leydig cells surrounded by tubules in stages IX to XII of the seminiferous epithelium cells are more heavily labeled (Gizang-Ginsberg and Wolgemuth, 1985).

Ontogeny of Gene Expression in the Fetal Rat Pituitary

A major application of *in situ* hybridization will be to localize specific mRNAs in early embryogenesis. Reliable methods for screening early embryonic sections for specific mRNAs will be especially useful in determining the cell types in which oncogenes, homeo-box containing genes, and growth factor genes, as well as other mRNAs that may play important roles in embryogenesis are expressed. Further, *in situ* hybridization should be able to demonstrate the stages at which specific genes become active in development and thus more precisely identify when developmentally important cellular interactions are occurring.

Thus, we have begun to investigate the feasibility of using *in situ* hybridization techniques to study the onset of gene expression in development. We have initially focused on the onset and location of POMC gene expression in the developing rat pituitary gland. It is known that in the adult rat pituitary gland, all cells in the intermediate lobe, and about 5% of the cells of the anterior lobe, produce POMC-derived peptides. We have begun to determine when during development POMC gene expression begins in these two different cell types and whether this coincides with, or is earlier than, the time these peptides are first detected immunocytochemically. This investigation was especially important because our studies have demonstrated that POMC proteolytic cleavages occurring at early stages of rat development are similar to those of the adult rat intermediate lobe. Thus, it was of interest to clarify whether anterior lobe cells, which at early stages of pituitary development are the first to accumulate immunocytochemically detectable POMC-peptides (Khachaturian et al., 1983; Nakane et al., 1977) are in fact the only ones that contain POMC mRNA at these stages. This would directly demonstrate that the extent of POMC proteolytic processing is changing in a specific population of cells during development.

Isolated embryos and fetuses e10 to e20 as well as neonates were fixed either by immersion or intracardiac perfusion with 4% paraformaldehyde, pH 7. Heads or isolated pituitaries were cut at 8 to 10 μm and hybridized with either ^{32}P-labeled cDNA probes or ^{3}H-labeled cDNA or cRNA probes. Subsequently, the slides were processed for either film or emulsion autoradiography, and adjacent sections were processed for β-endorphin immunoreactivity.

Since it is known that the relative abundance of POMC message is greater in the adult intermediate lobe than in the anterior lobe, we investigated when, during development, these differences in the relative abundance of POMC mRNA could first be detected by *in situ* hybridization. We initially approached this problem using ^{32}P-labeled probes. Rats e17 to adult were intracardially perfused, and the pituitaries were dissected, sectioned, and hybridized. In the adult, the intermediate lobe is much more heavily labeled than the anterior lobe. The same pattern is observed at early neonate stages (p17 and p2). Most important, even with this low level of resolution, heavier labeling over the intermediate lobe compared with the anterior lobe can be detected not only at e20 but as early as e17 (see Figure 10-3). These data suggest that even by e17 the relative concentration of POMC message is higher in the intermediate than in the anterior lobe.

^{3}H-labeled cDNA probes were then used to confirm and hopefully extend these observations to earlier developmental stages. Although these experiments confirmed the previous ^{32}P observations (the intermediate lobe of early postnatal and late fetal stages was more heavily labeled), specifically labeled anterior lobe cells could not be demonstrated prenatally, in part because of high backgrounds. We therefore began using cRNA POMC probes for these studies, which, as mentioned earlier, combine greater sensitivity with higher signal-to-noise ratios.

After hybridization of prenatal sections with ³H-POMC cRNA, we were able readily to detect single POMC mRNA–containing cells in the anterior lobe of the pituitary as early as e13.5 following 1 to 2 weeks of exposure of emulsion-coated sections. Most important, the distribution of radiolabeled cells and cells immunopositive for POMC peptides corresponds quite well, which suggests that we are detecting the onset of gene expression in both the anterior and intermediate lobes.

We were readily able to detect, following a 1-week exposure of e18 pituitary sections, a higher signal in the intermediate lobe than in the anterior lobe. This pattern correlated well with β-endorphin immunostaining of an adjacent section, in which almost all the cells of the intermediate lobe contain the peptide in addition to staining of some cells in the anterior lobe.

By e17, labeled cells above background in the intermediate lobe (not shown) were detected as well as specifically labeled cells in the anterior lobe (Figure 10-4). These observations correlated well with immunocytochemical data in which roughly 50% of the cells of the intermediate lobe contain β-endorphin at this age, in addition to the labeling of specific anterior lobe cells. Following a 2-week exposure of e16.5 sections, grain densities over the intermediate lobe well above the background levels in the posterior lobe and the underlying cartilage were observed and demonstrate that POMC mRNA can be detected by *in situ* hybridization in the intermediate lobe at the earliest times that we have demonstrated POMC-derived peptides immunocytochemically.

More important, at earlier stages of development (e.g., 13.5), the only POMC mRNA containing cells are located in the ventral region of the anterior lobe (Plate 7b), which corresponds precisely to the distribution of immunopositive cells at this age (Figure 10-5). Taken together with analysis of peptides present at these stages, these results suggest that POMC is being

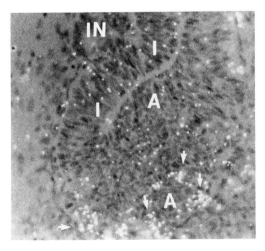

Figure 10-4 Hybridization of e13.5 rat pituitary with POMC cRNA. Sections were hybridized with POMC-118 cRNA, as before, and washed prior to RNase digestion in 0.2 × SSC (2 × 10 hours), RNase-treated (20 μg/ml at 37° C), and then washed to a final stringency of 0.2 × SSC at 42° C. Labeled cells are located only on the ventral part of the anterior lobe (A) (*arrows*), whereas the intermediate lobe (I) and infundibulum (IN) are unlabeled.

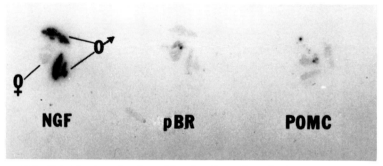

Figure 10-5   Hybridization of male and female submaxillary glands with NGF, POMC, and pBR DNA. Adjacent serial sections containing male and female submaxillary glands were hybridized with 20K cpm' of ^{32}P-labeled NGF cDNA, POMC cDNA, or pBR22 cDNA, as denoted in the figure, using the procedure described in Figure 10-2. After hybridization and washing, the slides were exposed to x-ray film for 2 days. Strong hybridization is seen in male submaxillary glands hybridized with ^{32}P-labeled NGF DNA, but this is not mimicked in sections hybridized with radiolabeled POMC and pBR DNA. Thus, the signal observed in male submaxillary glands results from specific hybridization of radiolabeled NGF DNA.

synthesized only in a subpopulation of anterior lobe cells at early ages and that the extent of POMC processing in the anterior lobe is changing as development proceeds. As a next step in this analysis, probes specific for intron regions of the POMC gene (which hybridize specifically to nuclei transcribing the POMC gene; Fremeau et al., 1986) will be hybridized to pituitary sections from early embryos to determine whether transcription of this gene during development significantly precedes its accumulation in the cytoplasm.

Hybridization of Growth Factor Probes

The expression of growth factor genes, such as NGF, EGF, and the insulinlike growth factors, are likely to be important during early development. We have successfully used NGF and EGF cDNA probes for *in situ* hybridization of the mouse submaxillary gland. We initially showed that *in situ* hybridization of male and female submaxillary glands with radiolabeled NGF probes reflects the difference in NGF mRNA content that is characteristic of this sexually dimorphic structure and demonstrates specific cell populations containing NGF and EGF mRNA.

We initially hybridized ^{32}P-labeled NGF to sections containing both male and female submaxillary gland, which can easily be distinguished histologically. Following the exposure of hybridized slides to x-ray film, film exposure (after 24 hours) was much greater over areas of the section containing male tissue (see Figure 10-5). This difference between male and female submaxillary gland was not seen if sections were hybridized with radiolabeled plasmid or POMC DNA (see Figure 10-5). These controls provide strong evidence that the differences seen reflect the known differences in NGF mRNA content of these tissues.

Further, we have hybridized submaxillary glands with ^{3}H-labeled NGF cDNA probes to obtain precise single-cell localization following emulsion autoradiography. Following a 4-week exposure, silver grains were concentrated in the male and female secretory tubules and were essentially absent in the acini; this cellular distribution of grains corresponds to the site of NGF storage in the submaxillary gland, as demonstrated by immunocytochemistry. These results show specific cellular localization of B-NGF mRNA in tissue sections using ^{3}H-labeled NGF probes and provide direct evidence that NGF is synthesized in secretory tubules. Strong hybridization signals are obtained following a 3-week exposure. Because an "average" eukaryotic cell contains approximately 2×10^5 mRNA molecules per cell, and the abundance of male submaxillary gland NGF mRNA is 0.1%, 200 copies of NGF mRNA are readily detected using double-stranded probes. To locate NGF mRNA containing cells in physiological sites of synthesis (where the abundance is about 1 copy per 10–100 cells) more sensitive procedures will be needed. Again, the use of cRNA has greatly improved the sensitivity; NGF mRNA in the secretory tubules is readily visualized in 2 days following hybridization with ^{35}S-labeled cRNA (Plate 7c) and 1 week following hybridization with ^{3}H-labeled NGF cRNA.

As mentioned earlier, previous applications of *in situ* hybridization have relied on other information about the cellular location of peptides of interest as a major criterion for the specificity of the hybridization. For example, in the demonstration that proopiomelanocortin mRNA was present in specific hypothalamic cells, adjacent sections were examined by immunocytochemistry and by *in situ* hybridization, and identical cells contained both immunoreactive endorphin and silver grains following binding of the POMC probe.

Recent immunocytochemical results have shown that NGF may be detectable in at least some of its cells in the adult animal where its synthesis has been suggested by blot analysis. However, alternative methods to ensure specificity of the *in situ* hybridization procedures for these growth factor gene probes must be developed, since immunocytochemical localization is controversial. In the procedure we are using to assess specificity, one nonspecific DNA probe and two specific but nonhomologous DNA probes are hybridized to three adjacent tissue sections. If similar grain distributions are seen in sections hybridized with the two different but gene-specific probes that are not mimicked in the section containing nonspecific DNA, the hybridization is considered specific. Following a short exposure of submaxillary gland tissue with nonhomologous EGF probes, the signal intensity of male tissue is much greater than that of female tissue, reflecting the size of the hybridization probes (Figure 10-6). It can be expected that use of such specificity controls in conjunction with hybridization using cRNA probes will allow the embryonic sites of synthesis of these developmentally important peptides to be determined.

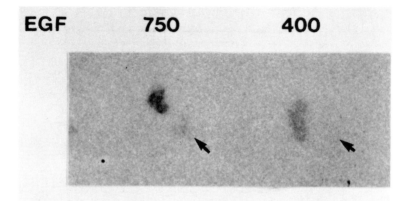

Figure 10-6 *In situ* hybridization of submaxillary glands with nonhomologous EGF cDNA fragments. As in Figure 10-5, male and female mouse submaxillary glands were fixed, embedded in the same block, and sectioned. Adjacent sections were hybridized with a ^{32}P-labeled 700-pb cDNA EGF fragment and a ^{32}P-labeled 400-bp EGF fragments (same protocol as in Figure 10-2), washed and exposed to x-ray film. Male and female tissue was identified following hematoxylin-eosin staining. Hybridization of both labeled nonhomologous EGF cDNA fragments is much greater over male tissue (*position of female tissue denoted by arrowhead*).

ACKNOWLEDGMENTS

The authors thank Jim Roberts, Beth Schachter, Cy Wilcox and Bob Fremeau, and Axel Ullrich for providing many of the cDNA and cRNA probes used in this work and for their continuing interest and advice. The research was supported in part by HD-18592 (JP), NS21970 (JP), and a Danforth-Compton Predoctoral Fellowship (DIL).

REFERENCES

Anderson, D.J. and Axel, R. Molecular probes for the development and plasticity of neural crest derivatives. *Cell* 42:649–662, 1985.

Angerer, L.M. and Angerer, R.C. Detection of poly A$^+$ RNA in sea urchin eggs and embryos by quantitative *in situ* hybridization. *Nucleic Acids Res.* 9:2819–2840, 1981.

Angerer, L., Deleon, D., Cox, K., Maxson, R., Kedes, L., Kaumeyer, J., Weinberg, E., and Angerer, R. Simultaneous expression of early and late histone messenger RNAs in individual cells during development of the sea urchin embryo. *Dev. Biol.* 112:157–166, 1985.

Blum, H.E., Haase, A.T., and Vyas, G.N. Molecular pathogenesis of hepatitis B virus infection: Simultaneous detection of viral DNA and antigens in paraffin-embedded liver sections. *Lancet* (Oct. 6):771–776, 1984.

Brahic, M. and Haase, A.T. Detection of viral sequences of low reiteration frequency by *in situ* hybridization. *Proc. Natl. Acad. Sci. USA* 75:6125–6129, 1978.

Brahic, M., Haase, A.T., and Cash, E. Simultaneous *in situ* detection of viral RNA and antigens. *Proc. Natl. Acad. Sci. USA* 81:5445–5448, 1984.

Brigati, D.J., Myerson, D., Leary, J.J., Spalholz, B., Travis, S.Z., Fong, C.K.Y., Hsiung, G.D., and Ward, D.C. Detection of viral genomes in cultured cells and paraffin-embedded tissue sections using biotin-labeled hybridization probes. *Virology* 126:32–50, 1983.

Brulet, P., Condamine, H., and Jacob, F. Spatial distribution of transcripts of the long repeated ETn sequence during early mouse embryogenesis. *Proc. Natl. Acad. Sci. USA* 82:2054–2058, 1985.

Cox, K.H., DeLeon, D.V., Angerer, L.M., and Angerer, R.C. Detection of mRNAs in sea urchin embryos by *in situ* hybridization using asymmetric RNA probes. *Dev. Biol.* 101:485–502, 1984.

Fremeau, R.T., Jr., Lundblad, J.R., Pritchett, D.B., Wilcox, J.N., and Roberts, J.L. Regulation of Pro-opiomelanocortin gene transcription in individual cell nuclei. *Science* 234:1265–1269, 1986.

Gee, C.E., Chen, C.-L.C., Roberts, J.L., Thompson, R., and Watson, S.J. Identification of proopiomelanocortin neurones in rat hypothalamus by *in situ* cDNA-mRNA hybridization. *Nature* 306:374–376, 1983.

Gee, C.E. and Roberts, J.L. Laboratory methods. *In situ* hybridization histochemistry: A technique for the study of gene expression in single cells. *DNA* 2:157, 1983.

Gizang-Ginsberg, E. and Wolgemuth, D.J. Localization of mRNAs in mouse testes by *in situ* hybridization: Distribution of a β-tubulin and developmental stage specificity of pro-opiomelanocortin transcripts. *Dev. Biol.* 111:293–305, 1985.

Griffin, W.S.T., Alejos, M., Nilaver, G., and Morrison, M.R. Brain protein and messenger RNA identification in the same cell. *Brain Res. Bull.* 10:597–601, 1983.

Haase, A., Brahic, M., Stowring, L., and Blum, H. Detection of viral nucleic acids by *in situ* hybridization. *Methods Virol.* VII:189–226, 1984.

Hafen, E., Levine, M., Garber, R.L., and Gehring, W.J. An improved *in situ* hybridization method for the detection of cellular RNAs in *Drosophila* tissue sections and its application for localizing transcripts of the homeotic *Antennapedia* gene complex. *EMBO* 2:617–623, 1983.

Hirose, S., Yamamoto, M., Kim, S.-J., Tsuchiya, M., and Murakami, K. Localization of renin mRNA in the mouse submandibular gland by *in situ* hybridization histochemistry. *Biomedical Res.* 4:591–596, 1983.

Jamrich, M., Mahon, K.A., Gavis, E.R., and Gall, J.G. Histone RNA in amphibian oocytes visualized by *in situ* hybridization to methacrylate-embedded tissue sections. *EMBO* 3:1939–1943, 1984.

Kelsey, J.E., Watson, S.J., Burke, S., Akil, H., and Roberts, J.L. Characterization of proopiomelanocortin mRNA detected by *in situ* hybridization. *J. Neuroscience* 6:38–42, 1986.

Khachaturian, H., Alessi, N.E., Munfakh, N., and Watson, S.J. Ontogeny of opioid and related peptides in the rat CNS and pituitary: An immunocytochemical study. *Life Sciences* 33:61–64, 1983.

Lawrence, J.B. and Singer, R.H. Quantitative analysis of *in situ* hybridization methods for the detection of actin gene expression. *Nucleic Acids Res.* 13:1777–1799, 1985.

Lewis, S.A. and Cowan, N.J. Temporal expression of mouse glial fibrillary acidic protein mRNA studied by a rapid *in situ* hybridization procedure. *J. Neurochem.* 45:913–919, 1985.

McGinnis, W., Levine, M.S., Hafen, E., Kuroiwa, A., and Gehring, W.J. A conserved DNA sequence in homoeotic genes of the *Drosophila* antennapedia and bithorax complexes. *Nature* 308:428–433, 1984.

Nakane, P.K., Setallo, G., and Merzurkiewicz, J.E. The origin of ACTH cells in rat anterior pituitary. *Ann. N.Y. Acad. Soc.* 297:201–204, 1977.

Pfeifer-Ohlsson, S., Goustin, A.S., Rydnert, J., Wahlstrom, T., Bjersing, L., Stehelin, D., and Ohlsson, R. Spatial and temporal pattern of cellular *myc* oncogene expression in developing human placenta: Implications for embryonic cell proliferation. *Cell* 38:585–596, 1984.

Pintar, J.E., Schachter, B.S., Herman, A.B., Durgerian, S., and Krieger, D.T. Characterization and localization of proopiomelanocortin messenger RNA in the adult rat testis. *Science* 225:632–634, 1984.

Rogers, A.W. *Techniques of Autoradiography*. 3rd Ed. New York: Elsevier North Holland, 1979.

Schwartzberg, D.G. and Nakane, P.K. *Endocrinology* 110:855–864, 1982.

Shivers, B.D., Harlan, R.E., Pfaff, D.W., and Schachter, B.S. Combination of immunocytochemistry and *in situ* hybridization in the same tissue section of rat pituitary. *J. Histochem. Cytochem.* 43:39–43, 1986.

Shivers, B.D., Schachter, B.S., and Pfaff, D.W. *In situ* hybridization for the study of gene expression in the brain. In *Methods in Enzymology*, Vol. 124:497–509, 1986.

Singer, R.H. and Ward, D.C. Actin gene expression visualized in chicken muscle tissue culture by using *in situ* hybridization with a biotinated nucleotide analog. *Proc. Natl. Acad. Sci. USA* 79:7331–7335, 1982.

Wilcox, J.N., Gee, C.E., and Roberts, J.L. *In situ* cDNA:mRNA hybridization: Development of a technique to measure mRNA levels in individual cells. *Methods in Enzymology*, Vol. 124:510–533, 1986.

11

Analysis of Viral Infections by *In Situ* Hybridization

Ashley T. Haase

VIRUSES AS MODEL SYSTEMS, AND THE RATIONALE FOR *IN SITU* HYBRIDIZATION

Viruses are obligate intracellular parasites that introduce a defined and relatively simple set of genes into cells. Expression of those genes in productive infections leads to a new crop of viruses that infect other cells. Although there may be no untoward effects of replication, more often, viruses kill, cripple, or transform cells. In whole organisms these pathological effects are limited by specific and nonspecific defense mechanisms that also usually eradicate the offending agent. The failure to eliminate an infectious agent results in a persistent infection, which may lead to cumulative pathological changes manifested months to years later. Infections of this kind are appropriately called slow infections.

In intact multicellular organisms, these events are generally widely separated in time nd   , and usually at or beyond the limits of population molecular analyses, which rely on macromolecular extracts from a population of cells in a tissue. When the infected cells of interest constitute only about 0.1 to 1% of the population, as is the case in humans and animals, we have used *in situ* hybridization (Haase et al., 1984) and, more recently, hybridization tomography (Haase et al., 1985b; Haase et al., 1985c) to overcome the problems of decreased sensitivity inherent in the use of pooled cellular nucleic acids.

This chapter is an account of the development and use of *in situ* hybridization in studies of the pathogenesis of viral infections. It begins with a detailed description of the methods that have proved successful in these investigations, and concludes with a brief recounting of the results that have been obtained to date on viral replication in brain and other tissues, the dissemination of virus in blood and other tissue fluids, the mechanism of tissue damage, and the potential role of viruses in chronic diseases of humans.

IN SITU HYBRIDIZATION: THE VISNA PARADIGM FOR QUANTITATIVE *IN SITU* HYBRIDIZATION

Visna is a paralytic condition of sheep (Brahic and Haase, 1981) that we have used as a model system to refine *in situ* hybridization methods for investigations of virus pathogenesis. The disease is caused by a retrovirus that multiplies in tissue culture cells to high titer in a few days. In this productive life cycle the RNA genome of the virus is reverse transcribed to ssDNA and dsDNA intermediates. The latter serves as template for transcription of thousands of copies of virus RNA that encode the virion proteins and are the sources of viral genomes of progeny viruses.

This tissue culture infection has been used to refine *in situ* hybridization methods to detect different species of viral nucleic acids (ssRNA and DNA, and dsDNA; Figures 11-1 and 11-2). At high multiplicities of infection, the vast majority of cells in the cultures participate in synthesis of viral nucleic acids so that population and *in situ* hybridization analyses can be performed on the same cultures. The population analyses provide information on the average number of copies per cell and calibrate and validate relationships to grain counts over a wide range of copy numbers (Figure 11-1). From the

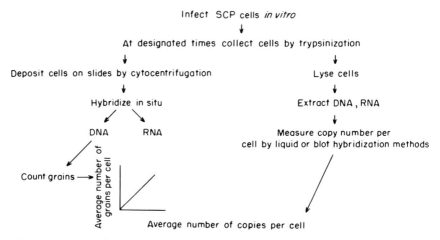

Figure 11-1 Flow diagram of procedures to quantitate synthesis of viral nucleic acids in single cells. Sheep choroid plexus cell cultures are infected at a moi of 3 PFU/cell. After adsorption of virus for 2 hours at 4° C, infection is initiated synchronously by the addition of warm medium. Cells are collected by trypsinization at specified times after infection, and the average content of nucleic acid per cell is determined in one aliquot. The remaining cells are sedimented onto slides where, after fixation, the content of nucleic acid in individual cells is assayed by hybridization *in situ*.

Visna DNA

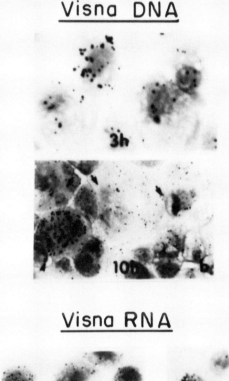

Visna RNA

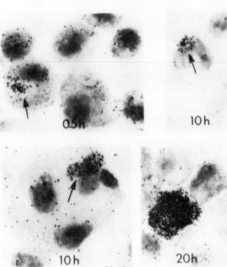

Figure 11-2 *In situ* hybridization measurements of visna DNA and RNA. Cells infected and collected, as described in Figure 11-1 were hybridized *in situ* for DNA and RNA. Heterogeneity of nucleic acid synthesis is evident at 10 and 20 hours (*arrows* indicate cells with low DNA synthesis; *arrowheads* indicate high DNA synthesis). For RNA, the arrow at 0.5 hours indicates entry of virus inoculum (approximately 30 copies of viral RNA). The arrow at 10 hours points to initiation of transcription in one nucleus. By 20 hours some cells have high copy numbers of RNA in nucleus and cytoplasm. (Used with permission of Haase et al., 1982.)

knowledge provided by the population analyses, we have been able to monitor the effects of changing the many variables that affect *in situ* hybridization. This information has been the basis for protocols that achieve hybridization efficiencies approaching 100% (Brahic and Haase, 1978), and sensitivities in the attogram (10^{-18} g) range (Blum et al., 1983), equivalent to a single copy of an average-sized mRNA. In parallel or subsequent stud-

ies, the general applicability of these methods has been examined with cellular RNAs (e.g., ^{3}H-uridine-labeled cells; cellular mRNAs) and other viral
DNAs and RNAs.

GENERAL CONSIDERATIONS: DIFFUSION, KINETICS, AND HYBRID STABILITY

The guiding principles for *in situ* hybridization of probes to nucleic acids in
cells are analogous to hybridization of probes to nucleic acids affixed to any
solid support (reviewed in Meinkoth and Wahl, 1984). In these situations,
the relevant parameters are diffusion of probe to target sequences, kinetics
of hybrid formation, and the thermal stability of the hybrid (Figure 11-3).

To enhance diffusion of probe, the hybridization reaction is driven by a
vast excess of probe in solution over target sequences in cells, by use of
probes in the range of 50 to 500 nucleotides (nt) in length for efficient entry,
and by treatments to permeabilize the cell and render target sequences more
accessible (e.g., detergents, acid extraction of basic proteins, and protease
digestion; Haase et al., 1984; Brahic and Haase, 1978).

The rate of hybrid formation with probe in vast excess follows first-order
kinetics and is primarily a function of the length and complexity (the number
of bases in a unique sequence) of the probe, temperature, and ionic strength.
(The contributions of pH and viscosity to the reaction rate are minor under
the hybridization conditions employed.) Hybrid stability is also governed by

Methodological Parameters for *in situ* Hybridization

– Fixation ⟨ Aldehydes / Precipitants

Compromise between diffusion and preservation of morphology ;
monitor retention of RNA, morphology, and hybridization efficiency.

– Diffusion of Probe to Target

Length and concentration of probe; temperature ; access (fixation,
permeability, accessibility e.g., detergents, HCl, protease).

– Kinetics of Hybrid formation – $t_{1/2} = \dfrac{N \ln 2}{3.5 \times 10^5 \cdot L^{0.5} \cdot C}$

N = complexity ; L = length; C = concentration

– Stability of hybrids –

$T_m = 81.5 + 16.6 \log M + 0.41 (\%G+C) - 820/L - 0.6 (\%F) - 1.4 (\%\text{mismatch})$

M = ionic strength; L = length ; F = formamide

Figure 11-3 Methodological parameters for *in situ* hybridization. See the text for explanation.

temperature, length, and ionic strength, and by additional factors such as base composition, concentration of denaturants like formamide that make it possible to conduct reactions at lower temperatures to preserve morphology, and mismatching of bases in probe and target. The relationship of these variables to rate and stability of hybrid formation are summarized in two equations (Meinkoth and Wahl, 1984; Thomas and Dancis, 1973):

$$t_{1/2} = \frac{N \ln 2}{3.5 \times 10^{5 \cdot L^{0.5}C}} \tag{1}$$

where $t_{1/2}$ = time (in seconds) required for half the probe to form hybrid
N = complexity (in nucleotides of unique sequence)
L = length of probe in nucleotides
C = concentration of probe in moles per liter
3.5×10^5 is the rate constant in 1 M Na$^+$

$T_m = 81.5°$ C $+ 16.6 \log M + 0.41(\%G + C)$
$$- \frac{820}{L} - 0.6(\%F) - 1.4(\% \text{ mismatch}) \tag{2}$$

where M = ionic strength (mol/liter)
F = formamide
T_m = temperature at which half of the hybrids are dissociated

Conditions for *in situ* hybridization are chosen to maximize the rates and stability of authentic hybrids, for example, by using probes of low complexity at high concentration. L is chosen as a compromise between increased hybridization with increasing length and decreased penetration. It should be noted that the actual rates of hybridization (Brahic and Haase, 1981) *in situ* are about an order of magnitude slower than predicted by equation (Haase et al., 1984). For visna RNA (N = 10,000), 2 days rather than hours are required to complete the reaction for a probe length of 50 to 500 nt (Brahic and Haase, 1978). For average-sized mRNAs (N = 2000) at a concentration of probe of 2×10^{-6} mol/liter (0.6 mg/ml) and L = 400 nt, the theoretical $t_{1/2}$ is about 5 minutes, whereas the observed $t_{1/2}$ was about 2 hours, which is in accord with the lower kinetics of *in situ* hybridization (Lawrence and Singer, 1985). With the increases in rates of reaction with polymers (e.g., dextran sulfate), hybridizations may be completed in a few hours to overnight for more complex nucleic acids.

The optimum temperature for hybridization (T_{opt}) is usually about T_m – 25° C (Meinkoth and Wahl, 1984). With M = 0.6 M Na$^+$, a base composition of 41% $G + C$, single-stranded probe of 50 nt in 50% formamide, the T_{opt} for *in situ* hybridization for visna would be 27° C, which is in reasonable agreement with the observation that ambient temperatures are convenient and close to optimal for *in situ* hybridization. Guidelines for shorter probes, such as oligonucleotides, are discussed in Meinkoth and Wahl (1984).

Most nick-translated probes are in the range of 200 to 500 nt, and the T_{opt} for *in situ* hybridization will be about 37° C.

FIXATION, EMPIRICAL ASPECTS OF *IN SITU* HYBRIDIZATION, AND ALDEHYDE REVERSAL PROTOCOLS

One of the paramount objectives of *in situ* hybridization, preservation of cellular morphology, is often at odds with sensitive detection of genes, which depends on diffusion of probe and efficient hybridization. From the standpoint of morphology, the aldehyde fixatives (glutaraldehyde, paraformaldehyde, and various combinations of the two with other fixatives such as alcohols or carbohydrate cross-linking fixatives; reviewed in Haase et al., 1984) may be better in some situations than the alcohols and other precipitant fixatives. However, hybridization efficiency may decrease by an order of magnitude or more as a consequence of decreased diffusion of probe, cross-linking protein to nucleic acids, and addition of methylol to the amino groups of nucleotides in the target nucleic acid, which prevents base pairing with probe sequences (Haase et al., 1984). These undesirable reactions, in turn, depend on the length of fixation, concentration of fixative, temperature and pH, penetrability of a particular tissue, and the secondary structure of nucleic acid. As a general rule, we have used and continue to recommend ethanol alone, or following use of ethanol/acetic acid, as the fixative of choice for cytocentrifuge preparations of frozen sections, at least at the initial stages of introducing *in situ* hybridization to a new system (Haase et al., 1984).

Relatively recently, various aldehyde fixatives have found increasing favor for *in situ* hybridization, with widely divergent recommendations as to fixation and hybridization conditions (Singer and Ward, 1982; Godard, 1983; Fournier et al., 1983; Hafen et al., 1983; McAllister and Rock, 1985). The often contradictory nature of these reports likely reflects the large number of interacting variables in *in situ* hybridization. Changing the conditions of fixation will also affect all the other parameters of hybridization. For each new system and fixative, the other variables of hybridization described in the preceding paragraphs must be reinvestigated to achieve the optimal balance between morphology and efficient hybridization. This empirical side of *in situ* hybridization is still guided by the general principles of solution-solid phase hybridization.

It is now possible to combine the excellent morphology of aldehyde fixation with high-efficiency *in situ* hybridization (Blum et al., 1984). In this method, the extent of aldehyde cross-linking is decreased before hybridization by protonating the methylene crossbridges or derivatized bases in HCl and triethanolamine. Using this method in conjunction with digitonin permeabilization and other diffusion enhancing pretreatments, hepatitis B viral nucleic acids were detected with less than a twofold decrease in sensitivity over frozen sections fixed in ethanol. Although the theory is untested, it seems likely from other systems that fixation for shorter times (20 minutes to 2 hours) and storage in alcohol (50–70%), to remove unreacted aldehydes and increase penetration of probe would increase hybridization further.

RADIOLABELED PROBES; POLYMER-MEDIATED INCREASES IN HYBRIDIZATION

The next consideration in *in situ* hybridization is the choice of probes (Figure 11-4)—radiolabeled or nonradioactive (Langer et al., 1981; Leary et al., 1983; Ruth, 1984; Forster et al., 1985; Landegent et al., 1985; Kempe et al., 1985; Chollet et al., 1985; Tchen et al., 1984), RNA (Cox et al., 1984) or DNA, single- or double-stranded. For maximum sensitivity and quantitation, radioactive probes are superior. In direct comparisons, these sensitivities exceed those obtained with biotinylated probes by about 2 orders of magnitude (unpublished). Nonradioactive probes, on the other hand, have the advantages of safety, speed, and convenience.

The number of copies of viral or cellular RNAs can be measured with single-stranded DNA (Haase et al., 1985c; Akam, 1983) or RNA probes (Cox et al., 1984), with double-stranded DNA probes (Haase et al., 1984; Haase et al., 1982), or with synthetic oligonucleotide probes (Haase et al., 1985a). In the current era these should be cloned probes, labeled by nick translation (dsDNA; Haase et al., 1984), by extension of a sequence primer (ssDNA probes in M13—Haase et al., 1985c; Akam, 1983; Haase et al., 1985a—or oligonucleotide—Uhl et al., 1985), or by transcription in the SP6 promotor vector system (ssRNA probes; Cox et al., 1984). As with fixation, a number of variables influence the outcome of hybridization. Single-stranded

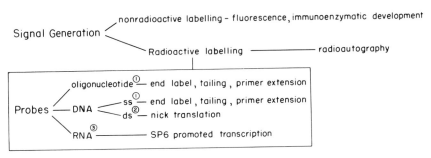

Probes for in situ Hybridization

Signal Generation
— nonradioactive labelling – fluorescence, immunoenzymatic development
— Radioactive labelling —————— radioautography

Probes — DNA — oligonucleotide ① — end label, tailing, primer extension
— ss ① — end label, tailing, primer extension
— ds ② — nick translation
— RNA ③ ————— SP6 promoted transcription

Variables in signal / noise

① High specific activities approaching 10^{10} dpm/μg; small target size
② Competing reassociation; polymer promoted increase in rate and extent of hybrid formation
③ Hybrid stability; reduced background with nuclease post treatment; decreased rate of hybrid formation; probe instability; shorter probes with specific activities of 10^9 dpm/μg

Figure 11-4 Considerations in choosing *in situ* hybridization probes. See the text for explanation.

RNA probes offer theoretical advantages of hybrid stability and better signal-to-noise ratios with ribonuclease posttreatments to reduce background (Cox et al., 1984). However, the rates of RNA–RNA hybrid formation are slower than those of DNA–RNA hybridizations (Casey and Davidson, 1977; Wetmur et al., 1981), and the length of the transcript decreases markedly with the low concentrations of labeled precursor needed for probes with specific activities of about 10^9 dpm/μg. This decrease in target size also applies to M13 ssDNA and to oligonucleotide probes (unpublished). For nick-translated ds probes, high specific ativities and complete representation of target sequences are combined, but the reassociation of the probe competes with hybrid formation (Haase et al., 1984; Haase et al., 1982). Sensitivities comparable to or even greater than single-stranded probes can nevertheless be achieved by polymer exclusion effects, which increase the rate and extent of hybrid formation and deposit aggregates of probe at each site of hybridization. Dextran sulfate, or polyethylene glycol (which, in our hands, gives lower backgrounds than dextran sulfate), increases hybridization 10- to 100-fold (Meinkoth and Wahl, 1984; Zimmerman and Harrison, 1985; Renz and Kurz, 1984). In deciding on an optimum probe, these complexities and the ease of shuttling cloned inserts between vectors make it worthwhile to compare several kinds of probes in a specific experimental system.

Probes for *in situ* hybridization will require labeling with ^{3}H, ^{125}I, or ^{35}S for good cellular resolution. Specific activities of 10^8 to 10^9 dpm/μg are readily obtained by the methods described in Haase et al. (1984; 1985c), Meinkoth and Wahl (1984), and Uhl et al. (1985). The efficiency of latent image formation for ^{3}H is 0.1 grain per disintegration, and 0.2 and 0.5 grain per disintegration, respectively, for ^{125}I and ^{35}S (Rogers, 1979).

BACKGROUND BINDING OF PROBE

Even highly specific cloned probes lacking vector sequences will bind nonspecifically, to some extent, to glass slides, cells, and other substrates. These nonspecific sites of adsorption can be blocked for the most part by treating the slides and cells with detergents, polymers such as heparin (Singh and Jones, 1984), Denhardts medium (DM: 0.02% each of bovine serum albumin, ficoll, and polyvinyl pyrrolidone), and by acetylating cells and slides (Haase et al., 1984). The latter step should be omitted for cellular RNAs and ssDNAs with negligible secondary structure.

Hybridization solutions also contain DM and usually poly-A and a nucleic acid unrelated to the probe (reduced to a similar size) to minimize nonspecific binding of probe to proteins, nucleic acids, and other molecules. Aurintricarboxylic acid (ATA), a potent inhibitor of protein-nucleic acid interactions (Gonzales et al., 1979), also improves backgrounds (Turtinen and Wietgrefe, unpublished).

The other important variables that affect the extent of nonspecific binding are the total radioactivity applied to the slide, posthybridization nuclease treatments, and washing. Generally, probe in at least 100-fold excess of target sequences is applied to the slide, limiting the total dpm per slide to 2×10^5 dpm. After RNA–RNA hybridizations, digestion with ribonuclease reduces background (Cox et al., 1984). Similar recommendations have been made for DNA–RNA hybridizations with a posthybridization digestion with S_1 nuclease (Godard, 1983). This step should be omitted in hybridizations in which nick-translated probes and polymers are used to form radioactive networks. The stringency, duration, and volume of the working procedures that follow hybridization depend on the length of radioautographic exposure. For short exposures, brief washes in hybridization medium, a moderately stringent wash in $2 \times$ SSC at $55°$ C (for visna, a stringent wash at $T_m - 12°$ C would be $60°$ C for a 50-nt probe), and a final wash in a liter of hybridization medium at room temperature for an hour suffice. Exceptionally low backgrounds (fewer than 5 grains per cell, for a 2-month exposure, with probe-specific activity of 5×10^8 dpm/μg) require more extensive washing at this last step.

DETAILED PROTOCOLS

In Situ Hybridization

Protocols that have proved reliable and sensitive in several viral systems are presented in Figure 11-5 as a flow diagram. Details of each step are set forth in the legend to the figure.

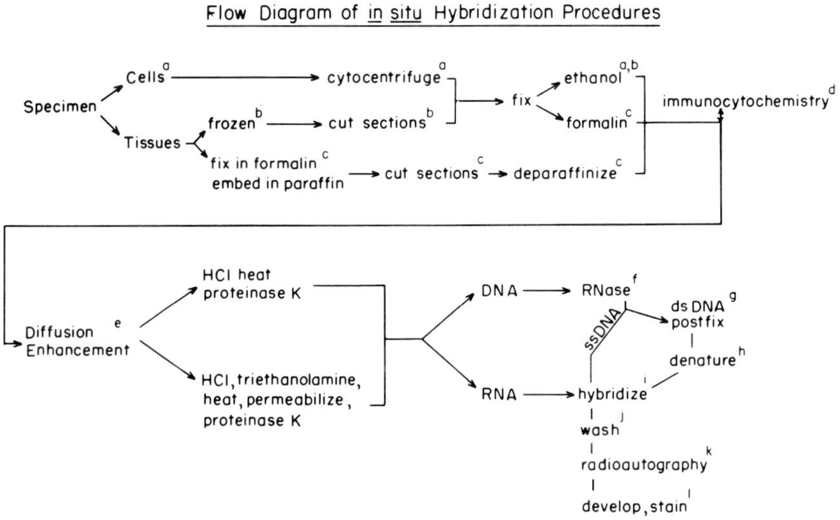

Flow Diagram of in situ Hybridization Procedures

Figure 11-5 *In situ* hybridization procedures. The figure presents a flow diagram of procedures. The small letters refer to footnotes with detailed protocols *(legend continued):*

Legend for Figure 11-5 *(continued)*

[a] Remove cells in monolayers with trypsin EDTA at 37° C, for 5 minutes. Centrifuge at 400 × g for 5 minutes. Resuspend in PBS lacking Ca^{2+} and Mg^{2+}. Recentrifuge and resuspend in PBS at 10^6 cells/ml. Cytocentrifuge at 400 rpm, for 5 minutes. Use washed and acetylated glass slides coated with DM. Air dry. Fix for 15 minutes in 3:1 ethanol/acetic acid, and for 5 minutes in ethanol at room temperature. Store desiccated at 4° C.

[b] Freeze 0.5- to 1.0-cm tissue sections between two blocks of dry ice. Store at −70° C. Cut into 4- to 10-μm sections. Air dry and fix in ethanol solutions or formalin for 20 minutes, as in c. Store in 70% ethanol.

[c] Perfuse animals with formalin solutions, or put tissue samples in 10% buffered formalin or (freshly prepared) 4% paraformaldehyde for 2 hours. Transfer and store at 4° C in 70% ethanol. Paraffin embed, cut into 5-μm section, and pick up the sections on treated glass slides coated with Elmer's glue. Alternatively, sprinkle gelatin on top of water at 42° C and pick the sections up; the gelatin will improve the adherence of the sections. Heat the slides at 37° C overnight and for 3 hours at 60° C. Deparaffinize in xylene (three 5-minute washes in xylene); wash twice in ethanol. Rehydrate through 90 and 70% ethanol and PBS before pretreating.

[d] Immunocytochemistry. Fix frozen sections in cold ethanol (4° C) for 20 minutes, or use aldehyde-fixed material. Rehydrate and permeabilize in cold PBS with 0.01% Triton X-100 for 4 minutes. Wash twice in PBS at 4° C, for 5 minutes each. React sequentially with normal serum diluted 1:200 in PBS (20 minutes, at room temperature); monospecific antisera diluted usually 1:100 to 1:300 in normal serum–PBS (30 minutes, at room temperature); 1:200 biotinylated antispecies IgG (30 minutes, at room temperature); 0.3% H_2O_2 in methanol for 30 minutes (to block endogenous peroxidase in monocytes, etc.), avidin biotinylated horseradish peroxidase complex (ABC reagent, Vector Labs) for 30 minutes; 0.2% H_2O_2 with 0.5 mg/ml phenylene-diamine HCl, 1 mg/ml pyrocatechol, or 0.5 mg/ml DAB in 0.1 M Tris HCl, pH 7.6, for 5 to 10 minutes at room temperature. Wash the slides in PBS for 5 minutes between steps and in water after reacting in DAB or PPD. Refix in ethanol, pretreat for *in situ* hybridization, and acetylate (0.25% acetic anhydride in triethanolamine buffer, pH 8.0) prior to hybridization. After development, stain with 0.5% $CuSO_4$, 1% methyl green as for color microradioautography.

[e] Place slides in 0.2 N HCl for 20 minutes, dip in water, or in 150 mmol/liter of triethanolamine for formalin-fixed material. Incubate the slides at 70° C in 2 × SSC (SSC = 0.15 mol/ liter of NaCl; 0.015 mol/liter of sodium citrate), dip in water, and permeabilize in 0.005% digitonin in 125 mmol/liter of sucrose, 60 mmol/liter of KCl, and 3 mM/liter of HEPES, pH 7.2. Dip the slides in water and transfer to a solution of 20 mmol/liter of Tris HCl, pH 7.4, 2 mmol/liter of $CaCl_2$, 5 μg/ml of a proteinase K. Digest at 37° C, for 15 minutes. Wash the slides twice in water, for 5 minutes each time, and dehydrate (for 5 minutes each, two washes in 70% ethanol and once in 95% ethanol), and air dry.

[f] For DNA measurements in cells that also contain viral RNA, the RNA must be removed by enzymatic treatment. Place the slides in 100 μg/ml of DNase-free ribonuclease A and 10 μg/ml of ribonuclease T_1 in 2 × SSC, at 37° C for 30 minutes. Digest in 20 μl under a coverslip in a humidified chamber. Remove the coverslip by washing the slide for 5 minutes in a freshly prepared solution of 0.1% diethylpyrocarbonate in 2 × SSC. Wash in 2 × SSC for 5 minutes and dehydrate (ssDNA), or postfix double-stranded DNA.

[g] Postfix dsDNA in freshly prepared paraformaldehyde. (Add 10 N NaOH dropwise to 5 g of paraformaldehyde, stirring in 100 ml of PBS. When the solution clears, neutralize with HCl and filter.) Place the slides in paraformaldehyde for 2 hours in the dark, wash twice in 2 × SSC, dip in water, and denature.

[h] Place slides in 95% formamide 0.1 × SSC at 65° C for 15 minutes. Quickly transfer to 0.1 × SSC at 4° C for 2 minutes, dip in water, and dehydrate.

[i] Concentrate probe by ethanol precipitation with 10 μg of DNA carrier used in the hybridization solution (e.g., calf thymus DNA reduced in size to about 300 by sonication and depurination followed by base treatment). Estimate the number of genes; use 0.3 ng of probe for 1 to 10 copies of a 3- to 10-kb gene per cell; 0.7 ng for 10 to 100 copies; 15 ng for

Legend for Figure 11-5 *(continued)*
1000 copies or more. The total radioactivity should not exceed 2×10^5 dpm/slide. Resuspend the probe (5 μl/slide) in hybridization solution containing 50% deionized formamide, 0.6 mol/liter of NaCl, 20 mmol/liter of HEPES, pH 7.2, 1 mM EDTA, 1 × DM, 100 μg/ml of calf thymus DNA, 50 μg/ml of poly-A, 5% PEG (~8000 molecular weight), and 100 μM ATA. Denature the probe by boiling it for 30 seconds and quench in ice. For ^{35}S probes add DTT to 10 mm. Add solution to cells or section, cover with a clean siliconized glass coverslip, and seal with rubber cement. Hybridize for 60 to 70 hours at 25° C to 37° C in the dark.

ʲRemove the coverslip by breaking the seal with the tip of a scalpel or needle. Wash the slides twice for 5 minutes in hybridization wash solution (HWS; 50% formamide, 0.6 mol/liter of NaCl, 1 mmol/liter of EDTA, 10 mM phosphate buffer, pH 6). Wash in 2 × SSC for 1 hour at 55° C. Wash twice at room temperature for 2 hours in 1 liter of HWS. Dehydrate in graded ethanols with 0.3 M ammonium acetate (to stabilize hybrids). For long exposure, wash for 3 days in 1.5 liter of HWS with scraps of nitrocellulose paper or Zetabind (Biorad).

ᵏIn total darkness, dip the dried slide in Kodak NTB-2 (or equivalent) nuclear track emulsion diluted 1:1 with 0.6 ammonium acetate. Dry for 2 hours, and place the slides in a light-tight box with a small amount of a drying agent. Wrap the boxes with aluminum foil and store at 4° C.

ˡWarm the boxes to room temperature for 1 hour. Develop in total darkness. Use Kodak D19 developer for 3 minutes at room temperature, dip in water, in fixer for 3 minutes, wash twice in water, 1 minute each time, then dry. Stain for 25 minutes in filtered Meyer's hematoxylin. Wash in distilled water for 15 minutes. Stain for 6 minutes in 0.25% eosin and 80% ethanol. Dehydrate through graded alcohols. Air dry. Leukocytes should be stained with May-Grunwald/Giemsa: Stain the RNA slides for 30 minutes in May-Grunwald. Stain the DNA slides for 45 minutes in May-Grunwald. (May-Grunwald is made up as a 0.3% solution in methanol, heated to 56° C for 1 hour, cooled to room temperature, and stirred overnight. Filter and dilute 1:1 with 0.01 M phosphate buffer, pH 6.8, to make a working solution.) Dip in phosphate buffer for 1 minute, dip in water, and stain for 45 minutes to 1 hour in Giemsa. (To make a working solution, dilute in phosphate buffer MCB concentrated stock 4–100 ml.) Decolorize in distilled water until the slides begin to turn from purple to blue. For color development of nuclear track emulsions, convert the black silver grains after development in D19 to magenta or cyan-colored grains as follows: harden the emulsion by immersing the slide for 3 minutes in 0.37% formalin buffered with 0.5% Na_2CO_3. Wash for 2 minutes in tap water at room temperature, and bleach for 1 minute in 10% $K_3Fe(CN)_6$, 5% KBr. Wash again in tap water, and develop for 1 minute in freshly prepared dye coupler. (For magenta grains, add 100 mg of Eastman Kodak M-38 to 5 ml of ethanol with 0.2 N NaOH. Mix with 45 ml of a solution containing Eastman Kodak developer CD2, 100 mg of Na_2SO_3, 50 mg of KBr, and 1 g of Na_2CO_3. For cyan grains, use Eastman Kodak cyan dye coupler C-16.) Wash, bleach, and wash again, and then fix for 5 minutes in $Na_2S_2O_3$ (24 g per 100 ml of water, 15 g of Na_2SO_3). Stain the cells in 0.5% $CuSO_4$ for 5 minutes, followed by a dip in water and 5 minutes in 1% methyl green. Wash briefly to remove excess dye. For double label *in situ* hybridization, coat the slides with a thin-barrier film made by dropping 0.4 ml of liquid Krylon (Borden), collected from a spray can, onto a water bath. Add the Krylon within a 40-cm circle of polyethylene tubing. As the solvent evaporates, a thin film of uniform thickness (silver interference color) forms in the center. Remove the tubing, and pick the film up from underneath with a developed and stained slide with magenta grains in the first layer. Dry in a semivertical position. Coat the Krylon film with a second layer of NTB-2 emulsion after "subbing" the film (0.5% gelatin, 0.05% CrK(SO$_4$)$_2$ · 12H$_2$O, for 5 minutes) to provide a substrate for the emulsion to bond to the film. Develop the second layer, and convert the grains to a cyan color.

Color Microradioautography and Double Label *In Situ* Hybridization

Color photographic development has recently been adapted to the microscopic scale for *in situ* hybridization (Haase et al., 1985e). Developed silver grains act as a microelectrode to oxidize developer:

1. developing agent $+ 2Ag^+X^-$ $2Ag^0 + 2H^+X^-$ + oxidized developer

The oxidized developer then reacts with a dye coupler to form a dye molecule.

2. oxidized developer + dye coupler intermediate (leucodye)
$$- 2AgX^- \text{ dye} + 2Ag^0 + 2H^+X^-$$

Because of the high pH (11.4) of these reactions, the gelatin in the nuclear track emulsion must be stabilized by cross-linking with formalin. The metallic silver is bleached at the beginning and the end of the oxidation reduction reactions.

Color development can be used with probes to different genes labeled with ^3H and ^{35}S to readily distinguish the probes. In this double label *in situ* hybridization technique, the first layer of NTB-2 is developed and the grains are converted to a magenta color. The second layer of emulsion over a thin clear film is developed, and grains are converted to a cyan color. ^3H-probes for one gene will produce magenta grains in the first layer, but will not penetrate to the second layer to contribute cyan grains. ^{35}S-probe for a second gene will produce magenta grains in the first layer, and cyan grains in the second. In this way, two different genes in the same cells or different cells in a section can be detected and quantitated (Plate 8).

COMBINED *IN SITU* HYBRIDIZATION AND IMMUNOLOGICAL METHODS

In situ hybridization can be combined with immunological assays to measure nucleic acids and proteins in the same cells (Brahic et al., 1984). The most convenient way to do this is to react sections first with antibody to a specific protein, develop the antibody–antigen complex with peroxidase and diaminobenzidene (DAB) or pyrocatecholphenylenediamine (PPD), and then hybridize. DAB or PPD forms insoluble polymers that withstand the hybridization procedures. The brown polymeric deposit serves as a marker for protein in the developed radioautograph, and silver grains over background provide evidence of a speicifc nucleic acid in the cell.

The immunologic reaction conditions must be optimized for fixation, dilution of antibody, and compatibility with *in situ* hybridization. Acetone, for example, provides good preservation of antigens, but gives poor results in the hybridization step. Fixation in ethanol alone, or in paraformaldehyde, is a good starting point. Antisera should be diluted as far as possible, because the level of hybridization decreases with increasing staining.

After the reaction with antibody and peroxidase-DAB or PPD, the slides must be acetylated to block nonspecific binding of probe to the polymer, particularly with DAB. Immunohistochemistry and subsequent *in situ* hybridization procedures are described in Figure 11-5.

COMBINED MACROSCOPIC AND MICROSCOPIC DETECTION OF VIRAL GENES

The principal limitation of *in situ* hybridization is its sampling power. This limitation can be surmounted without sacrificing the sensitivity and cellular resolution of *in situ* hybridization by using ^{125}I-labeled probes and sections of whole animals or whole organs (Haase et al., 1985b). The gamma emission of ^{125}I produces latent images on x-ray film at high efficiency in cassettes with calcium tungstate intensifying screens. The Auger electrons of ^{125}I produce latent images in nuclear track emulsions with good cellular resolution. The macroradioautograph provides a picture of the distribution of viral genes in a whole animal or organ. The nuclear track emulsion provides a picture of the cells that contain viral genes, located by using the x-ray image as a guide to the area of the section of interest. The sensitivity of the method can be increased by using probes labeled with both ^{125}I and ^{35}S for higher specific activities and efficiencies of grain development. Major steps and the principle of the method are summarized in Figure 11-6, and illustrations of the method are given in Figure 11-7. A detailed protocol follows:

1. Whole animals, organs, or large fragments of frozen tissues are embedded in 4% carboxymethyl cellulose (CMC) in a metal frame, the base of which attaches to an LKB 2258 cryomicrotome. After the metal frame is removed, the block is trimmed until the tissue is reached. A piece of 3M 336 tape is pressed securely over the block and tissue, and frozen sections of 15 to 30 μm are cut and transferred to the tape.
2. The section on the tape is air dried and fixed at 4° C in ethanol/acetic acid (15 minutes) and ethanol (5 minutes). The subsequent pretreatments to enhance diffusion of probe are those described for *in situ* hybridization: extraction with 0.2N HCl, heating at 70° C in 2 × SSC, and digestion with 1 μg/ml proteinase K. For DNA hybridizations, RNA is digested with ribonucleases, and the DNA is partially cross-linked with paraformaldehyde and denatured.
3. For hybridization, the tape is pressed section-side down on coated and acetylated glass slides (8.2 by 10.1 cm). Probe is introduced by puncturing the tape with a 25-gauge needle and injecting, between the tape and slide, 250 μl of buffered 50% formamide, 10% dextran sulfate containing 20 ng of probe (10^7 dpm). The needle hole is then sealed

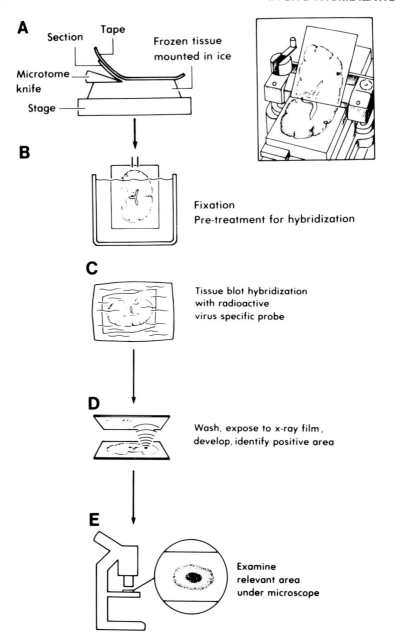

Figure 11-6 Combined macroscopic–microscopic hybridization. Principles and major steps in the method. (Reprinted with permission from Haase et al., 1985c.)

with rubber cement to complete formation of the hybridization chamber.
4. After hybridization for 72 hours, the tape is removed and secured to stainless steel frames (7.5 by 95.0 cm) that can be stacked and washed.

After washing, the sections are dehydrated and again placed section-side down against a plastic film (e.g., plastic bags and used for blot hybridizations). The hybridized sections are then placed with x-ray film in a cassette with two Cronex intensifying screens and exposed at $-70°$ C.

5. The tape is peeled off the plastic, coated with Kodak NTB-2 emulsion, dried, exposed, and later developed and stained. After trimming, the section is mounted by placing the section side up on mounting medium on a glass slide. More mounting medium is placed on the surface of the section, and a coverslip is rolled over the section. A weight is placed over the slip for 24 hours to create an optically flat surface. The areas of darkening in the x-ray film are used as a template to outline regions in the section that are to be examined microscopically, to identify cells with viral nucleic acid, and to distinguish hybridization signals from artifactual binding of probe.

CONTROLS

Because probes may bind to cellular constituents other than nucleic acids, and silver grains in the emulsion form in response to forms of energy other than radioactivity, it is essential to include a number of controls:

1. Chemography. Cellular constituents, dyes in stains, and other factors may induce latent images. Positive chemographic effects of this type should be excluded by coating a section with emulsion that has not been hybridized. If a positive chemographic effect is observed in the radioautograph, the stained section is coated after hybridization with a thin, clear film (e.g., Ixan, described in Rogers, 1979), prior to subbing and coating with NTB-2. Figure 11-8 illustrates an example of positive chemography.

2. Negative cells. The probe is hybridized to uninfected cells.

3. Nuclease control. The section is treated with ribonuclease before being hybridized for RNA, or with DNase (200 μg/ml ribonuclease-free DNase in 20 mmol/l Tris HCl, pH 7.4, 10 mm $MgCl_2$, at 37° C, for 60 minutes) before the postfixation step for detecting dsDNA. In most cases the hybridization should be reduced to the level in uninfected cells, although there are exceptions for interferon and other cellular mRNAs (Wietgrefe et al., 1985).

4. Heterologous probe. Hybridize subjacent sections with a probe to a gene that is not expected to be found in the tissue, such as vector DNA, which lacks cloned DNA insert, or visna probe in human cells. Hybridization should be equivalent to that in uninfected cells.

5. Reproducibility. A positive result should be obtained repeatedly in the sequential section of the same or related samples.

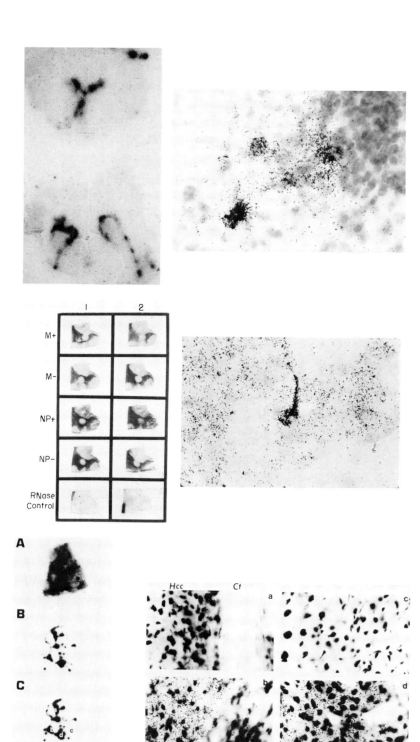

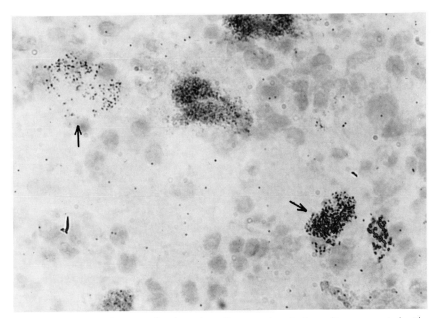

Figure 11-8 Positive chemography. A section of human trigeminal ganglion was coated with NTB-2 and developed and stained for several hours longer. Many of the large sensory neurons have large numbers of silver grains (*arrow*) over the nucleus and cytoplasm. This chemographic effect is probably induced by lipofuchsin and other pigments (D. Walker, A. Haase, unpublished).

INVESTIGATIONS OF VIRUS PATHOGENESIS BY *IN SITU* HYBRIDIZATION

The Visna Paradigm *In Vivo*

The replication of visna virus in animals stands in marked contrast to the productive infection of tissue culture cells used as a model system for de-

Figure 11-7 Combined macroscopic–microscopic hybridization. The panels on the left are the macroradioautographs of visna, measles, and hepatitis B virus infections (*top to bottom*); panels on the right show the corresponding microradioautographs. In the visna-infected and paralyzed sheep (*upper panels*), viral RNA is localized to paraventricular areas. Glial and neuronal cells at the margins of dense inflammatory infiltrates are infected. In subacute sclerosing panencephalitis, a slow infection caused by measles virus, the viral nucleocapsid protein (NP) and matrix (M) polypeptide genes and mRNAs are in neuronal and glial cells with inclusion bodies, and in cellular processes (*middle panels*). The wide distribution of viral genes in gray and white matter gives rise to the diffuse signal in the radioautographs at two levels (Haase et al., 1984; 1985a). In the lower panel, hepatitis B viral DNA sequences are localized primarily to infected liver cells (Ntl) that surround nodules of heptocellular carcinoma (Hcc). The tumor cells and connective tissue (Ct) in a and c lack detectable viral DNA (with a short exposure). Infected hepatocytes (b and d) produce the reproducible pattern in the macroscopic radioautographs (B and C). (D) A nuclease treated control. (A) A photograph of the specimen. (Modified from Haase et al., 1985b,c).

veloping and assessing *in situ* hybridization methods. In the infected animal, little virus is produced, and months or years elapse before the pathological effects of infection become evident. In this slow infection the virus persists and spreads despite a vigorous and sustained immune response by the host (Brahic and Haase, 1981).

Interest in understanding persistence of virus and the slow evolution of infection was the impetus for improving *in situ* hybridization methods. In our initial investigations of these central issues of pathogenesis, we found that hybridization of visna probes to DNA or RNA extracted from tissues failed to detect viral sequences; yet these same tissues readily produced virus on explantation. The conclusion from these studies that the viral genome was harbored in a covert state in a small proportion of the cells in a tissue gave us motive and rationale to improve *in situ* hybridization to detect and quantitate viral genes in the relatively rare infected cells in brain (Haase et al., 1977).

Restricted Gene Expression: The Persistence and Dissemination of Visna Virus and the Slow Evolution of Disease

With *in situ* hybridization we were successful in detecting viral genomes in about 1% of the cells in tissue sections (Haase et al., 1977). This result confirmed the highly focal nature of infection as the explanation for the negative results obtained by hybridization of visna probes to DNA extracted from the bulk population. Quantitation of viral RNA led to the discovery that the number of copies per cell was reduced by about 2 orders of magnitude compared with productive infection in tissue culture (Brahic et al., 1981). This restricted expression of viral genes in brain and in monocytes (Geballe et al., 1985; Peluso et al., 1985), and the accompanying block in synthesis of viral antigens, provided an explanation for the inability of the host's immune system to eradicate silently infected cells in tissues, and for the continued spread of virus in blood and tissue fluids with neutralizing antibody by a Trojan horse mechanism (the silently infected monocyte; Peluso et al., 1985). Figure 11-9 illustrates examples of *in situ* hybridization investigations of virus dissemination.

These studies also provided evidence of a spectrum of virus gene expression, and a correlation of the highest levels of gene expression with antigen production, inflammation, and tissue destruction (Stowring et al., 1985). These observations are consistent with an immunopathological process provoked and sustained by viral antigens. The slow appearance of disease is a consequence of the restriction in gene expression and the scanty production of viral antigens; and paralysis is the result of the destruction of cells that provide myelin in the nervous system, the oligodendrocyte (as shown by immunocytochemistry, cell-specific antisera, and *in situ* hybridization; Haase et al., 1981a).

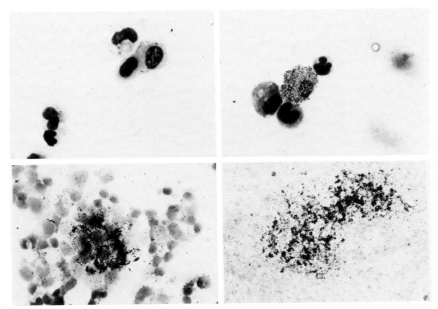

Figure 11-9 *In situ* hybridization studies of virus dissemination. The upper two panels demonstrate visna RNA in the nucleus of a binucleate macrophage, and in the nucleus and cytoplasm of a monocyte. These photographs were taken with polarized incident illumination so that the grains appear white. Monocytes harbor virus in the presence of neutralizing antibody, consisting with a Trojan horse mechanism of dissemination (Peluso et al., 1985). The lower left panel is a conventional photograph of a giant cell with measles virus RNA. The other cells are leukocytes from blood obtained from a patient with subacute sclerosing panencephalitis. The lower right panel is a low-power photomicrograph of a focus of lymphocytes with hepatitis B viral DNA localized at the margins of a blood vessel in the spleen of a patient with chronic active hepatitis.

Restricted Gene Expression as a General Theme in Slow Infections

In other slow infections, such as those caused by measles virus (Haase et al., 1985c; Haase et al., 1981a) Theiler's virus (Cash et al., 1985), mouse hepatitis virus (Lavi et al., 1984), and a paramyxovirus (Koch et al., 1984), similar constraints on the synthesis and expression of genomes are documented by quantitative *in situ* hybridization. Restricted gene expression in slow and persistent infections thus emerges as a general theme applicable to a taxonomically diverse group of RNA viruses as well as the DNA viruses, particularly those in the herpes virus family.

AN INDIGENOUS FLORA OF VIRAL GENES IN HUMANS

The realization that viral genes could be harbored in cells lacking detectable antigens or viral particles rekindled interest in the role of viruses in chronic diseases such as multiple sclerosis (MS). There is now evidence, from *in situ* and other hybridization methods, that paramyxoviruses, picornaviruses, and herpes simplex genomes can be detected in human brain, but in normal as well as MS and other diseases (Haase et al., 1981b; Haase et al., 1985d; Haase, in press). While the implications of these findings are still unclear, the discovery of viral genetic residue of earlier infections has enlarged our appreciation of the subtlety and longevity of virus host cell relationships. *In situ* hybridization and the combined macroscopic–microscopic technologies promise to broaden and deepen this understanding in the future and to help us evaluate the role of viruses in disease. This work and the studies of gene expression by *in situ* hybridization in the development, structure, organization, and function of normal cells and organisms are a compelling agenda for the future.

REFERENCES

Akam, M.E. The location of Ultrabithorax transcripts in Drosophila tissue sections. *EMBO J.* 2:2075–2084, 1983.

Blum, H.E., Stowring, L., Figus, A., Montgomery, C.K., Haase, A.T., and Vyas, G.N. Detection of hepatitis B virus DNA in hepatocytes, bile duct epithelium, and vascular elements by *in situ* hybridization. *Proc. Natl. Acad. Sci. USA* 80:6685–6688, 1983.

Blum, H.E., Haase, A.T., and Vyas, G.N. Molecular pathogenesis of hepatitis B infection: Simultaneous detection of viral DNA and antigens in paraffin-embedded liver sections. *Lancet* (Oct. 6):771–776, 1984.

Brahic, M. and Haase, A.T. Detection of viral sequences of low reiteration frequency by *in situ* hybridization. *Proc. Natl. Acad. Sci. USA* 75:6125–6129, 1978.

Brahic, M. and Haase, A.T. Lentivirinae: Maedi/visna virus group infections. Comparative aspects and diagnosis. In *Comparative Diagnosis of Viral Diseases*, Vol. IV, Part B. E. Kurstak and C. Kurstak (Eds.). New York: Academic Press, 1981, pp. 619–643.

Brahic, M., Stowring, L., Ventura, P., and Haase, A.T. Gene expression in visna virus infection. *Nature* 292:240–242, 1981.

Brahic, M., Haase, A.T., and Cash, E. Simultaneous *in situ* detection of viral RNA and antigens. *Proc. Natl. Acad. Sci. USA* 81:5445–5448, 1984.

Casey, J. and Davidson, N. Rates of formation and thermal stabilities of RNA:DNA and DNA:DNA duplexes at high concentrations of formamide. *Nucleic Acids Res.* 4:1539–1552, 1977.

Cash, E., Chamorro, M., and Brahic, M. Theiler's virus RNA and protein synthesis in the central nervous system of demyelinating mice. *Virology* 144:290–294, 1985.

Chollet, A. and Kawashima, E.H. Biotin-labeled synthetic oligodeoxyribonucleotides: Chemical synthesis and uses as hybridization. *Nucleic Acids Res.* 13:1529–1541, 1985.

Cox, K.H., DeLeon, D.V., Angerer, L.M., and Angerer, R.C. Detection of mRNAs in sea urchin embryos by *in situ* hybridization using asymmetric RNA probes. *Dev. Biol.* 101:485–502, 1984.

Forster, A.C., McInnes, J.L., Skingle, D.C., and Symons, R.H. Non-radioactive hybridization probes prepared by the chemical labelling of DNA and RNA with a novel reagent, photobiotin. *Nucleic Acids Res.* 13:745–761, 1985.

Fournier, J.-G., Rozenblatt, S., and Bouteille, M. Localization of measles virus nucleic acid sequences in infected cells by *in situ* hybridization. *Biol. Cell* 49:287–290, 1983.

Geballe, A.P., Ventura, P., Stowring, L., and Haase, A.T. Quantitative analysis of visna virus replication *in vivo*. *Virology* 141:148–154, 1985.

Godard, C.M. Improved method for detection of cellular transcripts by *in situ* hybridization: Detection of virus-specific RNA in Rous sarcoma virus-infected cells by *in situ* hybridization to cDNA. *Histochemistry* 77:123–131, 1983.

Gonzalez, R.G., Blackburn, B.J., and Schleich, T. Fractionation and structural elucidation of the active components of aurintricarboxylic acid, a potent inhibitor of protein nucleic acid interactions. *Biochim. Biophys. Acta* 562:534–545, 1979.

Haase, A.T., Stowring, L.S., Narayan, O., Griffin, D., and Price, D. Slow persistent infection caused by visna virus: Role of host restriction. *Science* 195:175–177, 1977.

Haase, A.T., Swoveland, P., Stowring, L., Ventura, P., Johnson, K.P., Norrby, E., and Gibbs, Jr., C.J. Measles virus genome in infections of the central nervous system. *J. Infect. Dis.* 144:154–160, 1981a.

Haase, A.T., Ventura, P., Gibbs, C.J., and Tourtellotte, W.W. Measles virus nucleotide sequences: Detection by hybridization *in situ*. *Science* 212:672–675, 1981b.

Haase, A.T., Stowring, L., Harris, J.D., Traynor, B., Ventura, P., Peluso, R., and Brahic, M. Visna DNA synthesis and the tempo of infection *in vitro*. *Virology* 119:399–410, 1982.

Haase, A.T., Brahic, M., and Stowring, L. Detection of viral nucleic acids by *in situ* hybridization. In *Methods in Virology*, Vol. VII, K. Maramorosch and H. Koprowski (Eds.). New York: Academic Press, 1984, pp. 189–226.

Haase, A.T., Blum, H., Stowring, L., Geballe, A., Brahic, M., and Jensen, R. Hybridization analysis of viral infection at the single-cell level. In *Clinical Laboratory Molecular Analyses*. New York: Grune & Stratton, 1985a.

Haase, A.T., Gantz, D., Blum, H., Stowring, L., Ventura, P., Geballe, A., Moyer, B., and Brahic, M. Combined macroscopic and microscopic detection of viral genes in tissues. *Virology* 140:201–206, 1985b.

Haase, A.T., Gantz, D., Eble, B., Walker, D., Stowring, L., Ventura, P., Blum, H., Wietgrefe, S., Zupancic, M., Tourtellotte, W., Gibbs, Jr. C.J., Norrby, E., and Rozenblatt, S. Natural history of restricted synthesis and expression of measles virus genes in subacute sclerosing panencephalitis. *Proc. Natl. Acad. Sci. USA* 82:3020–3024, 1985c.

Haase, A.T., Stowring, L., Ventura, P., Burks, J., Ebers, G., Tourtellotte, W., and Warren, K. Detection by hybridization of viral infection of the human central nervous system. *Ann. N.Y. Acad. Sci.* 436:103–108, 1985d.

Haase, A.T., Walker, D., Stowring, L., Ventura, P., Geballe, A., Blum, H., Brahic, M., Goldberg, R., and O'Brien, K. Detection of two viral genomes in single cells by double-label hybridization *in situ* and color microradioautography. *Science* 227:189–192, 1985e.

Haase, A.T. *In situ* hybridization and covert virus infections. In *Concepts in Viral Pathogenesis II*. A.L. Notkins and M.B.A. Oldstone (Eds.). New York: Springer-Verlag, in press.

Hafen, E., Levine, M., Garber, R.L., and Gehring, W.J. An improved *in situ* hybridization method for the detection of cellular RNAs in Drosophila tissue sections and its application for localizing transcripts of the homeotic Antennapedia gene complex. *EMBO J.* 2:617–623, 1983.

Kempe, T., Sundquist, W.I., Chow, F., and Hu, S-L. Chemical and enzymatic biotin-labeling of oligodeoxyribonucleotides. *Nucleic Acids Res.* 13:45–57, 1985.

Koch, E.M., Neubert, W.J., and Hofschneider, P.H. Lifelong persistence of paramyxovirus Sendai-6/94 in C129 mice: Detection of a latent viral RNA by hybridization with a cloned genomic cDNA probe. *Virology* 136:78–88, 1984.

Landegent, J.E., Jansen in de Wal, N., van Ommen, G.-J.B., Baas, F., de Vijlder, J.J.M., van Duijn, P., and van der Ploeg, M. Chromosomal localization of a unique gene by non-autoradiographic *in situ* hybridization. *Nature* 317:175–177, 1985.

Langer, P.R., Waldrop, A.A., and Ward, D.C. Enzymatic synthesis of biotin-labeled polynucleotides: Novel nucleic acid affinity probes. *Proc. Natl. Acad. Sci. USA* 78:6633–6637, 1981.

Lavi, E., Gilden, D.H., Highkin, M.K., and Weiss, S.R. Persistence of mouse hepatitis virus A59 RNA in a slow virus demyelinating infection in mice as detected by *in situ* hybridization. *J. Virol.* 50:563–566, 1984.

Lawrence, J.B. and Singer, R.H. Quantitative analysis of *in situ* hybridization methods for the detection of actin gene expression. *Nucleic Acids Res.* 13:1777–1799, 1985.

Leary, J.J., Brigati, D.J., and Ward, D.C. Rapid and sensitive colorimetric method for visualizing biotin-labeled DNA probes hybridized to DNA or RNA immobilized on nitrocellulose: Bio-blots. *Proc. Natl. Acad. Sci. USA* 80:4045–4049, 1983.

McAllister, H.A. and Rock, D.L. Comparative usefulness of tissue fixatives for *in situ* viral nucleic acid hybridization. *J. Histochem. Cytochem.* 33:1026–1032, 1985.

Meinkoth, J. and Wahl, G. Hybridization of nucleic acids immobilized on solid supports. *Anal. Biochem.* 138:267–284, 1984.

Peluso, R., Haase, A.T., Stowring, L., Edwards, M., and Ventura, P. A Trojan horse mechanism for the spread of visna virus in monocytes. *Virology* 147:231–236, 1985.

Renz, M. and Kurz, C. A colorimetric method for DNA hybridization. *Nucleic Acids Res.* 12:3435–3444, 1984.

Rogers, A.W. *Techniques of Autoradiography*. 3rd Ed. Amsterdam: Elsevier North Holland, 1979, p. 308.

Ruth, J.L. Chemical synthesis of non-radioactively-labelled DNA hybridization probes. *DNA* 3:123, 1984.

Singer, R.H. and Ward, D.C. Actin gene expression visualized in chicken muscle tissue culture by using *in situ* hybridization with a biotinated nucleotide analog. *Proc. Natl. Acad. Sci. USA* 79:7331–7335, 1982.

Singh, L. and Jones, K.W. The use of heparin as a simple cost-effective means of controlling background in nucleic acid hybridization procedures. *Nucleic Acids Res.* 12:5627–5638, 1984.

Stowring, L., Haase, A.T., Petursson, G., Georgsson, P., Palsson, P., Lutley, R., Roos, R., and Szuchet, S. Detection of visna virus antigens and RNA in glial cells in foci of demylination. *Virology* 141:311–318, 1985.

Tchen, P., Fuchs, R.P.P., Sage, E., and Leng, M. Chemically modified nucleic acids as immunodetectable probes in hybridization experiments. *Proc. Natl. Acad. Sci. USA* 81:3466–3470, 1984.

Thomas, Jr., C.A. and Dancis, B.M. Ring stability. Appendix to: Formation of rings from Drosophila DNA fragments. *J. Mol. Biol.* 77:44–55, 1973.

Uhl, G.R., Zingg, H.H., and Habener, J.F. Vasopressin mRNA *in situ* hybridization: Localization and regulation studied with oligonucleotide cDNA probes in normal and Brattleboro rat hypothalamus. *Proc. Natl. Acad. Sci. USA* 82:5555–5559, 1985.

Wetmur, J.G., Ruyechan, W.T., and Douthart, R.J. Denaturation and renaturation of *Penicillium chrysogenum* mycophage double-stranded ribonucleic acid in tetra-alkylammonium salt solutions. *Biochemistry* 20:2999–3002, 1981.

Wietgrefe, S., Zupancic, M., Haase, A., Chesebro, B., Race, R., Frey II, W., Rustan, T., and Friedman, R.L. Cloning of a gene whose expression is increased in scrapie and in senile plaques in human brain. *Science* 230:1177–1179, 1985.

Zimmerman, S.B. and Harrison, B. Macromolecular crowding accelerates the cohesion of DNA fragments with complementary termini. *Nucleic Acids Res.* 13:2241–2249, 1985.

Appendix A: Discussion

At the symposium that spawned this book, each of the contributing authors presented a 1-hour talk about his or her current work involving *in situ* hybridization technology. At the end of each talk there were questions from the audience, which brought to light many of the concerns of people attempting to set up and use this technique in their laboratories. Because the answers to these questions are of general interest, we have summarized them in this appendix.

1. What is the nature of some of the common artifacts encountered in doing *in situ* hybridizations? (Discussants: Griffin, Haase)

A certain class of silver grains can sometimes be subjectively identified as artifacts on the basis of how they appear. These are the result of positive chemography, which may be difficult to control for. Since chemographic artifacts do not result, in general, from the method of tissue preservation, more than likely they are the result of some contaminant in the prehybridization or hybridization buffers. One component that is notoriously unpredictable is dextran sulfate. The batch-to-batch variation is such that sometimes it gives a good signal and other times the background is tremendously high. Additionally, some of the staining procedures used to visualize cell morphology and architecture also give positive chemographic signals—methylene blue in particular. Hematoxylin-eosin stains do not seem to cause as much of a problem. One way around this problem is to stain the sections postemulsion development. The only way to be entirely sure that the signal one is seeing is authentic specific hybridization is to do the proper controls—for example, DNasing or RNasing of probe, use of heterologous probes, RNasing of tissue section, using unlabeled probe, and so on.

2. What is an appropriate signal-to-noise ratio to ensure that one's signal is specific? (Discussants: Haase, Schachter)

This ratio can vary anywhere from 10 to 100:1. The key to this problem is the level of background, which depends to a great extent on the specific activity of the probe one is using. Datum that helps in determining this number is the knowledge of the abundance of that particular RNA in the tissue section one is examining. Such calculations require a knowledge of tissue thickness, amount of specific RNA in the tissue (from Northern blot analysis), and relative distribution of the signal in different cell types represented in the tissue section. Knowing these facts as well as the efficiency of autoradiographic grain development per disintegration (^{3}H—0.1 grain/disintegration; ^{125}I—0.3 grains/disintegration; and ^{35}S—0.5 grains/disintegration) enables one to determine the hybridization efficiency. Generally, the best idea for determining the proper signal-to-noise ratio is to do the experiment with and without competitor to see how the signal changes.

3. Does permeabilization of the tissue section with different detergents affect the autoradiographic signal? (Discussant: Haase)

In most systems, Triton X-100 works fine, yet the simplest and morphologically most satisfactory detergent is digitonin (0.005%). If one is using detergent to help permeabilization, one should start with digitonin and vary its concentration as well as the type of detergent, as dictated by the signals achieved.

4. How does dextran sulfate affect *in situ* hybridization signals when RNA probes are used? (Discussants: Angerer, Haase, Higgins)

When the hybridization experiment is done, simultaneously, with and without 10% dextran sulfate, the signal is enhanced fivefold in the presence of dextran sulfate. This enhancement persists over a 72-hour hybridization period. There are probably two reasons for this: first, the increase in hybridization rate; second, that dextran sulfate can serve as a nonspecific competitor for the binding of RNA probes to tissue, thus essentially increasing the amount of probe capable of hybridizing to the tissue section RNA. This is true whether the glass slides are polylysine coated or gelatin subbed. Another advantage to using dextran sulfate is that the viscosity it affords actually spaces the coverslips, thus controlling the depth of the hybridization solution as well as keeping the coverslip in place.

5. What are some of the cautions to be taken in using RNase to reduce backgrounds when using RNA probes? (Discussant: Angerer)

If the RNase step is performed in 0.5 M NaCl, there is no loss of the double-stranded (specific) signal. Another factor that may be important is the type of fixation procedure used to keep the RNA in the tissue section. Lightly fixed tissues, such as those fixed with ethanol/acetic acid, may be more susceptible to loss of hybridization signal than tissues fixed with glutaraldehyde, with which every RNA molecule is probably fixed at many points along its length.

6. Must one use radioactively labeled probes immediately after they are synthesized? (Discussants: Angerer, Haase, Singer)

The only thing to be concerned about is the shortening of probes upon long-term storage because of degradation induced by radiative decay. When these shortened probes become smaller than 30 nucleotides, the appropriate hybridization temperature can change drastically from that required for a probe 100 bases in length. One can store "maximally" labeled ^{3}H-probes for months with little change in their hybridization characteristics. To check probe degradation, one can run a gel to determine if the probe size is still appropriate for the desired hybridization conditions.

7. What is an appropriate size for RNA probes to be used in *in situ* studies? (Discussants: Angerer, Singer, Watson)

Probes of 50 to 100 bases in length give signals at least three times higher than probes that are 500 bases long.

8. What are some of the differences between hybridization of nucleic acids in solution and hybridization of a freely soluble nucleic acid to one bound to a solid support, such as RNA in fixed tissue? (Discussants: Angerer, Haase, Singer, Watson)

There are many variables to take into account in making this comparison. The smaller the probe, the better it diffuses into a tissue section, thus making the *in situ* hybridizatiton relatively faster for this probe over a longer probe. This order of change is not reflected in the same way in solution hybridization in which penetration is not a factor. The kinetics of *in situ* hybridization reactions have not been well worked out; still, we can use the technique and empirically change variables to maximize our signal.

9. When using biotinylated probes, does the degree of biotinylation affect the hybridization signal? (Discussant: Singer)

Probes that have more than 20 molecules of biotin per kilobase do not give an increased detection signal compared with probes containing less biotin. In fact, the rigidity of a probe labeled with many molecules of biotin may make it difficult for the probe to diffuse into a tissue section, and hence may reduce the detectable signal. Additionally, since biotin is lipidophilic, having a large number of biotins on a probe may cause it to partition into areas of the cell that are rich in lipids, such as membranes. Another possible reason for the occasional difficulty in detecting biotin-labeled probes, is that lipids may bind to the biotin molecule, making it difficult for a reporter molecule to attach to the biotin. These problems are currently being investigated.

10. Which is more sensitive in detecting a biotinylated probe, peroxidase or alkaline phosphatase? (Discussant: Singer)

Alkaline phosphatase is 10 times better at detecting a biotin molecule than peroxidase. One problem with alkaline phosphatase detection in some tissue sections, notably the brain, is the high level of endogenous alkaline phosphatase, which increases the background.

11. Do hybridization times have to be increased for biotinylated probes versus isotopically labeled probes? (Discussants: Singer, Watson)

The nucleic acid saturation curve for biotinylated probes seems to level out at approximately the same time in cells and tissues as it does for isotopically labeled probes, given the same conditions.

12. What is the best way of determining that the hybridization signal is specific to the probe that one is using? (Discussants: Angerer, Haase, Pintar, Schachter, Watson)

There are three ways currently being used to address this question. One way, which is probably the easiest, is to do a competition study on adjacent slides; if the signal is specific, it should specifically compete away. A second way of looking into this problem is to use probes derived from different regions of the cDNA or gene of interest to probe adjacent tissue sections; if the probe is specific, the same hybridization pattern should be generated with the different probes. The third way is to isolate total RNA from the fixed tissue. This RNA can then be used to do a Northern blot to show that the size of the message being detected with the probe is the same as that from nonfixed tissue. Unfortunately, this latter technique can be used only for mRNAs of relatively high abundance.

13. Do nonspecific DNAs used in prehybridization and hybridization buffers, such as salmon sperm DNA, alter the hybridization signal of a specific probe? (Discussants: Angerer, Griffin, Singer, Watson)

The only way such "background blockers" could alter the signal would be for those DNAs to have some homology with the specific probe being used. Even if this were the case, the homologous gene would have to be represented in more than single-copy numbers to be effective. It might be possible that if the nonspecific DNA is too large or present in too high a concentration, steric hindrance of hybridization or precipitation of probe DNA or RNA onto the tissue section may result.

14. There are many different recipes in common use for making hybridization buffer, which one is the best? (Discussants: Angerer, Haase, Schachter, Watson)

The best advice is to select one hybridization buffer recipe and work out the variables of one's system using that protocol. Generally, the critical thing about these buffers is that given the salt concentration and length of probe one is using, one should do the hybridization at 25 degrees below the melting temperature. To increase the stringency, one can lower the salt concentration and increase the temperature during the washing steps. If this is done, most of the differences in the various hybridization buffers will even out, leaving an acceptable signal-to-noise ratio.

15. In doing combined immunohistochemistry and *in situ* hybridization in the same brain section, does colchicine treatment, used to increase peptide levels in the cell body so that the peptide is easily detectable, alter mRNA levels? (Discussants: Higgins, Schachter, Watson)

Colchicine treatment does not seem to alter the *in situ* hybridization signal.

16. Which should be done first—*in situ* hybridization or immunohistochemistry on sections where one wants to combine both techniques? (Discussants: Haase, Griffin, Schachter, Watson)

It is best to do the hybridization first followed by immunohistochemistry. If immunohistochemistry is done first, followed by the hybridication protocol, the peroxidase or perhaps the reaction product of the peroxidase is lost, so that there is very little immunohistochemical signal.

17. Does proteinase K or other tissue permeabilizers actually help increase the *in situ* hybridization signal? (Discussants: Angerer, Haase, Schachter, Watson)

This depends on how one prepares their tissue for the *in situ* technique. In most cases in which the tissue is fixed and then sectioned, the permeabilization step is necessary. In other cases, in which fresh-frozen tissue is used, the method of freezing the block may be such that the ice crystals formed during freezing actually serve to permeabilize the tissue, so that additional permeabilizers are not necessary.

18. In doing immunohistochemistry and *in situ* in the same section, does one have to worry about RNase being present in the antibody solution? (Discussants: Schachter, Watson)

Endogenous RNase activity can, indeed, reduce one's *in situ* signal. To eliminate this as a problem, one should treat one's antibody solution with an RNase inhibitor of some sort. In lieu of purifying the antibody away from contaminating RNase activity by affinity purification, one could add an aliquot of the vanadyl-ribonucleoside RNase inhibitor to the antibody, or perhaps even titrate out the RNase by adding RNA to the antibody solution. Surprisingly, contrary to what might be expected, 0.2% diethylpyrocarbonate does not inactivate some antibodies' binding abilities; hence it can be added to an antibody solution to inhibit any RNase activity. Of course, the appropriate control experiments should be performed before one actually tries this on precious tissue sections.

19. What are some of the considerations to keep in mind in preparing radioactively labeled probes for *in situ* hybridization studies? (Discussants: Angerer, Haase, Higgins, Singer, Watson)

When preparing single-stranded probes such as those made from M13 templates, one can generally get full-length transcripts (assuming 1000 bases is full length) at a specific activity of 10^7 disintegrations per minute (dpm) per microgram of transcript. To get higher specific activities, one usually uses less labeled nucleotide—perhaps an order of magnitude below the K_m of the enzyme—producing shorter transcripts of higher specific activity. One solution to this problem is to add higher concentrations of the radioactive precursor. This can become quite expensive, but perhaps more important, the shipments of radioactivity sometimes contain an inhibitor that can inactivate DNA polymerase (also SP6 polymerase can be inactivated by this inhibitor). Unfortunately, it is impossible to predict which batches will contain the inhibitor and which will not.

For S^{35}-labeling of riboprobes where backgrounds have been reported to be a problem, it is important to perform all of the hybridization and washing steps in reducing conditions. This can be achieved by adding dithiothreitol (DTT) to all of the solutions used in these steps. The higher the concentration of DTT (up to 300 mmol), the lower the background. A reasonable, and not too expensive, concentration to use is 100 mmol. The point at which DTT is no longer necessary is the RNasing step, in which high levels of DTT may inhibit its enzymatic activity. Again, different lots of ^{35}S-labeled nucleotide, as shipped from the manufacturer, vary in the level of background signal obtained.

An important consideration in using the SP6 vector system to make riboprobes is to make sure that the restriction enzyme used to linearize the template has cut the DNA completely. This is critical because there is a DNA sequence in the vector, 3′ to the multiple cloning site, which will bind to ribosomal RNA. If the template is not completely truncated, high levels of binding to ribosomal RNAs can occur. Complete truncation can usually be achieved by cutting the DNA twice with an excess of the same enzyme. If this is not satisfactory, the truncated DNA can be purified from the uncut material by electrophoresis and the desired DNA can be isolated by electroelution.

20. When using a variety of tissues, is there a need to alter fixation conditions to optimize the *in situ* signal? (Discussants: Coghlan, Haase, Higgins, Schachter, Watson)

When using the ethanol/acetic acid (3:1) fixation method, if an ethanol wash follows the ethanol/acetic acid treatment there is generally little need to alter the fixation conditions. The morphology, hybridization efficiency, and *in situ* signal achieved using this technique are excellent for all tissues, with the possible exception of brain tissue. These fixatives, when used in brain, will remove myelin and as a result may give poor morphology. When examining brain tissue, it is probably wise to use an aldehyde fixative because they provide better retention of morphology.

Another consideration is whether or not immunohistochemistry will be performed on the same sections; if so, the precipitating fixatives such as ethanol/acetic acid may not be sufficient to fix high enough quantities of the antigen in the cell to give a good signal. One reagent to avoid in fixing one's tissue is acetone because it is incompatible with *in situ* hybridization protocols.

The fact that aldehyde fixatives can be used as a method of preserving tissues for *in situ* hybridization studies allows one the possibility of using sections provided by the Brain Bank of McLean Hospital. This is critical in the study of certain mental disorders because fresh tissue from affected individuals is scarce and it may be possible to use formalin-fixed tissue that has been stockpiled over the past several years. In addition, tissue preserved in paraffin blocks (as available from pathology departments) is stable for years as long as the tissue was fixed and embedded quickly after dissection.

In storing tissue sections for use in *in situ* hybridization studies, it is preferable to transfer sections from aldehyde-containing solutions to 70% ethanol. This transfer has two results: it helps in permeabilizing the tissue for probe hybridization, and it removes unreacted aldehydes. If these sections are stored at 4° C they should be stable for years. It is also possible to store tissue sections, after alcohol dehydration, air drying, and dessication at 4° C for months.

Appendix B
Approaches to the Study of Gene Expression in the Monkey Cerebral Cortex by Means of *In Situ* Probes

E. G. Jones, S. H. C. Hendry, P. Isackson, and R. A. Bradshaw

It is now becoming clear that the expression of neurotransmitter-related enzymes and that of neuromodulatory peptides in nerve cells of the peripheral nervous system and in neuroendocrine tissues such as the adrenal medulla are activity dependent (see Black et al., 1984; LaGamma et al., 1985). Depolarization of sympathetic ganglion cells by synaptic activity *in vivo* or under artificial conditions *in vitro* clearly leads to induction of the catecholamine-synthesizing enzyme, tyrosine hydroxylase, by increasing the abundance of its mRNA, while interruption of impulse activity by denervation leads to a great increase in substance P levels in the same cells. In the adrenal medulla, denervation leads to a decrease in levels of tyrosine hydroxylase and of another catecholamine-synthesizing enzyme, phenylethanolamine N-methyl transferase, but to increased levels of leu-enkephalin and increased abundance of the mRNA for the leu-enkephalin precursor molecule.

These observations, which suggest fundamental mechanisms whereby neuronal phenotype (as expressed by transmitter status) may be modulated and whereby neuronal transmitter levels may be regulated, have led us to search for related phenomena in the central nervous system, particularly in the cerebral cortex.

We have made two sets of primary observations in the monkey cerebral cortex that lend themselves to further analysis from the preceding points of view. The first set of data relates to the fact that certain populations of neurons in the cerebral cortex can express both a classic transmitter, such as acetylchloine (ACh) or gamma-aminobutyric acid (GABA) and one or more of the better known neuropeptides. The first became evident in 1984.

At that time, Eckenstein and Baughman described colocalization of immunoreactivity for choline acetyl transferase (ChAT), the synthesizing enzyme for ACh, and for vasoactive intestinal polypeptide (VIP) in about 80% of VIP-positive neurons of the rat cortex. At the same time, three groups reported on the colocalization of GABA or glutamic acid decarboxylase (GAD) immunoreactivity with that for one or more neuropeptides (Schmechel et al., 1984; Somogyi et al., 1984; Hendry et al., 1984) (Figure B-1). In our studies on the monkey cortex, we now find that 100% of cholecystokinin (CCK) immunoreactive cells, 95 to 97% of somatostatin (SRIF) cells, 95 to 97% of neuropeptide Y (NPY) cells, and 90% of substance P (or other tachykinin) cells colocalize GABA. Furthermore, NPY and SRIF are almost invariably colocalized, and a small number of the putative SP immunoreactive cells also colocalize NPY.

It is important to point out, however, that these colocalizations occur in only about 20% of the total GABAergic cell population and seem to involve one major cell class (typically a long, stringy, bipolar, or bitufted form; Figure B-2) (Jones et al., 1986). The remaining GABA cells, including several well-characterized morphological forms, never display immunoreactivity for a known peptide. This first set of data has led us to inquire if all GABA neurons transcribe the message for the GABA-synthesizing enzyme but simply do not produce sufficient GAD or GABA to reach threshold for

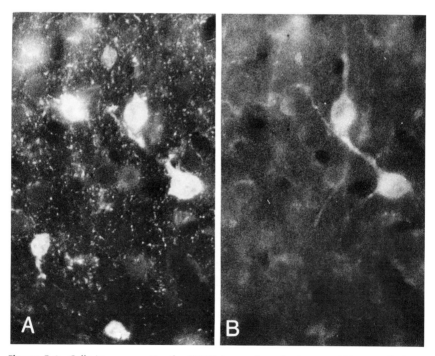

Figures B-1   Cells immunoreactive for GABA in monkey visual cortex (A), two of which prove, when excitation filters are switched, to also be immunoreactive for CCK (B).

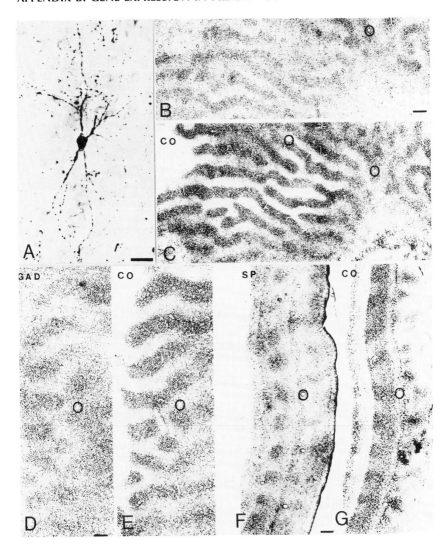

Figure B-2 (A) A cell from the monkey motor cortex immunoreactive for NPY and typical of most of those that colocalize GABA and many known cortical neuropeptides. (B, C) Alternative tangential sections of monkey visual cortex, stained for cytochrome oxidase (CO, B) and GABA, showing reduction in GABA immunoreactivity in ocular dominance columns deprived of pattern vision by monocular eyelid suture for 11 weeks. (D, E) Alternative sections from the same monkey showing reduction in GAD immunoreactivity in deprived eye columns. (F, G) Alternative vertical sections from another monocular-deprived adult monkey, showing reduction of SP immunoreactivity in the deprived eye columns.

immunocytochemical detection. We also ask whether or not cells coexpress-
ing GAD and one or more peptides regulate this expression by differential
transcription.

The second set of data relevant to the points made earlier relate to the
effect of monocular visual deprivation on immunoreactivity for GABA, GAD,
SP, and a particular protein kinase in the visual cortex of adult monkeys
(Hendry et al., 1985; Hendry and Jones, 1986; Hendry and Kennedy, 1986).
After more than 9 weeks of monocular visual deprivation resulting from
suturing the lids of one eye shut (or 1 week following actual removal of the
eye), alternating deprived and nondeprived ocular dominance columns can
be revealed in the primary visual cortex by staining for the mitochondrial
enzyme, cytochrome oxidase (see Figure B-2). Concomitantly, in the deprived columns the number of cells and cell processes that can be stained
immunocytochemically for GAD, GABA, and SP fall dramatically. No cells
actually die, however; the amounts of these compounds apparently fall below threshold levels for immunocytochemical detection (Figure B-2). At the
same time, immunoreactivity for a calcium/calmodulin-dependent protein
kinase (CaM II kinase), actually increases in the deprived columns (Hendry
and Kennedy, 1986). These observations lead us to ask if the reduction in
synaptic driving to cells in the deprived columns leads to the differential
expression of the compounds mentioned through differential regulation of
transcription of the genes for GAD, for preprotachykinin, and for CaM II
kinase.

METHODS

Synthetic 30-base oligonucleotide probes have been prepared in a nucleotide
synthesizer, based on cDNA sequences published in the literature for the
known prepropeptides for human SP, NPY, SRIF, and CCK. These have
been 5'-end labeled with gamma ^{32}P (Maniatis et al., 1982). A full-length
cDNA probe for cat GAD, provided by Dr. Alan Tobin (Kaufman et al.,
1986), has been expressed in E. coli and labeled with ^{32}P by nick translation
(Maniatis et al., 1982). All of these probes have now been used for in situ
hybridization to sections from monkey cerebral cortex, essentially following
the protocol of Watson et al. (see Chapter 7). The best results were obtained
from brains fixed by perfusion with weak aldehyde mixtures and sectioned
frozen at 10 μm. Following hybridization, autoradiography was performed
using Kodak NTB$_2$ emulsion with exposure times of 3 to 7 days.

RESULTS AND DISCUSSION

The results are preliminary, and it will be some time before we have answers
to the several questions posed here. At the moment it is possible only to

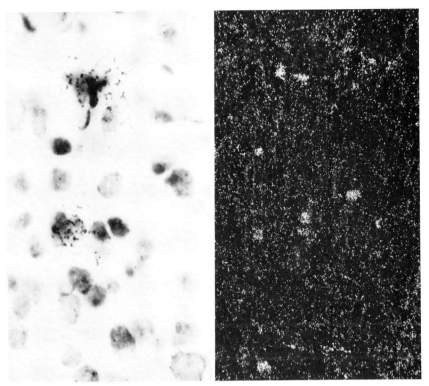

Figure B-3   (*Left*) *In situ* hybridization of a ^{32}P-labeled, synthetic oligonucleotide probe complementary to the precursor for human NPY. The probe has hybridized to two cells in this field from monkey visual cortex. Other cells, stained with thionin, are unlabeled. (*Right*) Lower-magnification darkfield photomicrograph of an adjacent field.

state that the method of *in situ* hybridization can be used to conduct meaningful experiments on central nervous tissue in which tissue structure and cell morphology is sufficiently preserved to make accurate histological localization, cell identification, and cell counting feasible. Up to this point, our results indicate that the message for the precursors of those compounds we have thus far been able to localize is contained in a limited number of the total population of cortical neurons (Figure B-3). The proportion of cells showing evidence of the mRNA for a particular peptide precursor, for example, appears very similar to the proportion that are immunoreactive for that particular peptide. In the cat also, Kaufman et al. (1986) have shown that the message for GAD is invariably colocalized in cells that display GAD immunoreactivity. It thus appears that cells not normally expressing a particular neuroactive substance, its precursor, or its synthetic enzyme(s) are not transcribing the genes for these molecules.

Our experiments have not reached the stage at which we can say whether monocular visual deprivation affects transmitter, peptide, or enzyme levels by affecting levels of mRNA in the cells of the deprived eye columns. If this should prove to be so, it may be necessary to resort to quantitative autoradiography to detect changes in abundance of the message. It will then probably be necessary to resort to probes labeled with lower-energy beta-particle emitters, such as ^{35}S to reduce dispersion of the signal.

ACKNOWLEDGMENTS

This work was supported by grants NS 21377 and NS 1964 from the National Institutes of Health, U.S. Public Health Service.

REFERENCES

Black, I.B., Adler, J.E., Dreyfus, C.F., Jonakait, G.M., Katz, D.M., LaGamma, E.F., and Markey, K.M. Neurotransmitter plasticity at the molecular level. *Science* 225: 1266–1270, 1984.

Eckenstein, F. and Baughman, R.W. Two types of cholinergic innervation in cortex, one co-localized with vasoactive intestinal polypeptide. *Nature* 314: 153–155, 1984.

Hendry, S.H.C. and Jones, E.G. Reduction in number of immunostained GABA neurons in deprived-eye dominance columns of monkey area 17. *Nature* 320: 750–753, 1986.

Hendry, S.H.C., Jones, E.G., De Felipe, J., Schmechel, D., Brandon, C., and Emson, P.C. Neuropeptide containing neurons of the cerebral cortex are also GABAergic. *Proc. Natl. Acad. Sci. USA* 81: 6526–6530, 1984.

Hendry, S.H.C., Jones, E.G., and Kennedy, M. Modulation of GABA, substance P and protein kinase immunoreactivities in monkey striate cortex following eye removal. *Neurosci. Abstr.* 11: 16, 1985.

Hendry, S.H.C. and Kennedy, M. Immunoreactivity for a calmodulin-dependent protein kinase is selectively increased in macaque striate cortex after monocular deprivation. *Proc. Natl. Acad. Sci. USA* 83: 1536–1540, 1986.

Jones, E.G., Hendry, S.H.C., and De Felipe, J. GABA-peptide neurons of the primate cerebral cortex: A limited cell class. In *Cerebral Cortex, Vol. VI.* E.G. Jones and A. Peters (Eds.). New York: Plenum Press, 1986.

Kaufman, D.L., McGuinis, J.F., Krieger, N.R., and Tobin, A.J. Brain glutamate decarboxylase cloned in λgt-11: Fusion protein produces γ-aminobutyric acid. *Science* 232: 1138–1140, 1986.

LaGamma, E.F., White, J.D., Adler, J.E., Krause, J.E., McKelvy, J.F., and Black, I.B. Depolarization regulates adrenal preproenkaphalin mRNA. *Proc. Nat. Acad. Sci. USA* 82: 8252–8255, 1985.

Maniatis, T., Frisch, E.F., and Sambrook, J. *Molecular Cloning, A Laboratory Manual.* New York: Cold Spring Harbor Laboratory, 1982.

Schmechel, D.E., Vickrey, B.G., Fitzpatrick, D., and Elde, R.P. GABAergic neurons of mammalian cerebral cortex: Widespread subclass defined by somatostatin content. *Neurosci. Lett.* 47: 227–232, 1984.

Somogyi, P., Hodgson, A.J., Smith, A.D., Nunzi, M.G., Gorio, A., and Wu, J.-Y. Different populations of GABAergic neurons in the visual cortex and hippocampus of the cat contain somatostatin- or cholecystokinin immunoreactive material. *J. Neurosci.* 4: 2590–2603, 1984.

Appendix C:
List of Attendees

Dr. Mary Abood
University of California
San Francisco

Dr. Ann Arvin
Stanford University

Dr. Guriq Basi
Stanford University

Dr. Klaus G. Bensch
Stanford University

Dr. Helen M. Blau
Stanford University

Dr. Sarah Bodary
University of California
San Francisco

Dr. Jerold Chun
Stanford University

Dr. Roland Ciaranello
Stanford University

Dr. Dennis Clegg
University of California
San Francisco

Dr. Christian Deschepper
University of California
San Francisco

Dr. Karen Downs
University of California
San Francisco

Dr. Dea Eisner
Stanford University

Dr. Justin Fallon
Stanford University

Dr. Peter Gebicke-Haerter
Stanford University

Dr. Mark Hamblin
Stanford University

Dr. Phyllis Hanson
Stanford University

Dr. Edna Hardeman
Stanford University

Dr. Stewart Hendry
University of California Irvine

Dr. Raymond L. Hintz
Stanford University

Dr. Andrew Hoffman
Stanford University

Dr. Richard Jacobson
Stanford University

Dr. Alexandra Joyner
University of California
San Francisco

Dr. Judy Kalinyak
Stanford University

Dr. Barry Kaplan
University of Pittsburgh

Dr. Michael LaBate
Stanford University

Dr. Ping Law
University of California
 San Francisco

Dr. Nancy Lee
University of California
 San Francisco

Dr. Wally Lind
The Salk Institute

Dr. Alan Louie
University of California
 San Francisco

Dr. Robert Margolskee
Stanford University

Dr. Michael McPhaul
Stanford University

Dr. Synthia Mellon
University of California
 San Francisco

Dr. Christophe Mueller
Stanford University

Dr. Marc Navré
Stanford University

Dr. Elly Nedivi
Stanford University

Dr. David Newell
Stanford University

Dr. Craig Okada
Stanford University

Dr. Donna Peehl
Stanford University

Dr. Andrew J. Perlman
Stanford University

Dr. Carol Phelps
University of Rochester

Dr. Ron Rosenfeld
Stanford University

Dr. Kevin Roth
Washington University

Dr. David Shelton
Stanford University

Dr. Thomas Sherman
University of Michigan
 School of Medicine

Dr. Robert Shiurba
Veterans Administration Hospital

Dr. Richard Simerly
The Salk Institute

Dr. J. H. Pate Skene
Stanford University

Dr. Robert Thompson
Oregon Health Sciences University

Dr. Susie Torti
Stanford University

Dr. Cy Wilcox
Genentech Inc.

Dr. Dona Wong
Stanford University

Index

Acetic anhydride, tissue preparation and, 14, 36, 50
Acetylcholine, cerebral cortex and, 231
Acquired immune deficiency syndrome, HTLV-III and, 20
Actin mRNA, solution hybridization and, 81
Actin sequences, detection of, 80
Adenosine triphosphate. *See* ATP
Adrenal medulla, denervation and, 231
AIDS. *See* Acquired immune deficiency syndrome
Albumin, slide preparation and, 74
Alcohols, deproteination and, 10
Aldehyde fixatives. *See specific type*
Alkaline phosphatase, biotinylated, mRNA detection and, 87–88, 223–224
Amygdala, 154, 156–158
Anterior commissure, 172–173
Antibiotin antibody, hybrid detection and, 20, 85
Antibiotin-colloidal gold complexes, *in situ* hybridization and, 20
Antigens
    peptide hormone, *in situ* hybridization and, 18
    viral
        detection of, 116
        in situ hybridization and, 18
Arginine-vasopressin probe, hybridization and, 133
Astrocytes, glial fibrillary acidic protein and, 164
ATP, oligonucleotide probe labeling and, 14
Autoradiography, hybrid detection and, 60–62, 84, 89
Avidin, hybrid detection and, 19, 85–88

Base mismatches, hybridization rate and, 8
Biotinyl-dUTP, biotin labeling and, 19

Bouin's fixative, 35, 97–100, 114
Bovine serum albumin, tissue preparation and, 15
Brain-specific gene 1B236, 164–168

Carnoy's fixative, 10, 35, 73, 114
cDNA probes
    chicken beta-actin, 72
    disadvantages of, 11
    double-stranded
        denaturation of, 15
        posthybridization treatments and, 17
        single-stranded vs., 185–186
    epidermal growth factor, 192–193
    hybridization temperatures and, 15
    labeling of, 13, 27
    nerve growth factor, 192–193
    nick translation and, 11, 13, 27, 36, 156
    sources of, 129–130
cDNA-RNA hybrids, stability of, 130
cDNA subtractive hybridization, 150–151, 156–158
Cell culture, *in situ* hybridization and, 90
Cerebral cortex, 154, 231–236
Cholecystokinin immunoreactive cells, GABA colocalization and, 232
Choline acetyltransferase, immunoreactivity for, colocalization of, 232
Chromosomal DNA, *in situ* hybridization and, 82–84
Chromosomes
    fixation protocol for, 74–75
    metaphase, *in situ* hybridization and, 82, 91
Colchicine, *in situ* hybridization and, 225
Color microradioautography, 207–208
Complementary DNA probes. *See* cDNA probes
Complementary RNA probes. *See* cRNA probes